Patriarchy and Class

African Modernization
and Development Series

Paul Lovejoy, Series Editor

Patriarchy and Class

African Women in the Home and the Workforce

EDITED BY

Sharon B. Stichter
and Jane L. Parpart

Westview Press

BOULDER & LONDON

African Modernization and Development Series

Copyright © 1988 by Westview Press, Inc.

Published in 1988 in the United States of America by Westview Press, Inc., 5500 Central Avenue, Boulder, Colorado 80301

Library of Congress Cataloging-in-Publication Data
Patriarchy and class.
  (African modernization and development series)
  Includes index.
  1. Women—Africa, Sub-Saharan—Social conditions.
2. Sex role—Africa, Sub-Saharan. 3. Patriarchy—
Africa, Sub-Saharan. 4. Africa, Sub-Saharan—Social
conditions—1960–   . I. Stichter, Sharon B.
II. Parpart, Jane L. III. Series.
HQ1787.P29 1988   305.4'0967   88-219
ISBN 0-8133-7416-2

Printed and bound in the United States of America

The paper used in this publication meets the requirements of the American National Standard for Permanence of Paper for Printed Library Materials Z39.48-1984.

6   5   4   3   2   1

Contents

Acknowledgments

This book has a long history and many debts. The idea for a collection or special issue on marxist-feminist thought as applied to Africa dates back to the 1983 Boston meetings of the African Studies Association, when the editors organized a panel with a number of the book's contributors, including Luise White, Nancy Folbre, and Jeanne Henn. The panel aroused considerable interest and raised the possibility of a special journal issue on the subject. But the decision to produce a collection only crystallized the following spring at the annual meetings of the Canadian Association of African Studies in Montreal. At the time, Paul Lovejoy had just become the editor of Sage's series on African Modernization and Development, and he enthusiastically endorsed the suggestion that the proposed collection would be appropriate for the series. The authors gratefully acknowledge his support and help throughout the project, especially during the series' smooth transition from Sage to Westview Press. His thoughtful comments facilitated revisions and made *Patriarchy and Class* a better book.

Above all, we wish to thank our contributors, who have produced fine original essays addressing the book's major themes and have willingly responded to suggestions for revision. Earlier versions of some of these essays appeared in the publications of the African Studies Center at Boston University and have consequently benefitted from the careful editing of its production editor, Jean Hay.

We also wish to thank the staff at Westview Press, particularly Sally Furgeson and Dean Birkenkamp, the external reviewers for their insightful comments, and our typists Mary Wyman and Becky Grant, both of the Dalhousie University History Department. We also wish to thank Dalhousie's faculty of graduate studies and CIDA's public participation programme for financial support towards manuscript preparation costs. The book could not have been completed without them. Finally, our grateful thanks to our husbands, Joe Stichter and Tim Shaw, without whose patience, loving support, and endurance through countless hours of restaurant conversations this book could surely have never been completed.

Sharon B. Stichter
Jane L. Parpart

1

Introduction: Towards a Materialist Perspective on African Women

Sharon B. Stichter and Jane L. Parpart

The great debate between feminism and marxism, so central to the development of contemporary feminist theory, seems in many ways to have passed African studies by. Marxist analyses focusing on modes of production, capital accumulation and class formation in Africa have abounded, yet most of these have paid scant attention to feminist concerns. On the other hand, empirical studies of the position of African women have flourished, yet the implications of these for a marxist-feminist approach[1] have not been drawn out.

The present volume argues for the applicability of a materialist mode of production analysis to the situation of women in Africa. Although the contributors by no means agree on precise theoretical formulations, nor even on the use of the term "mode of production," they are united in addressing themselves in one way or another to fundamental problems of conceptualization relevant to marxism-feminism. Some of the chapters are couched in a specifically theoretical vein, while others have as their primary aim the interpreting of particular experiences of women in concrete African social formations. This introduction will briefly review some of the intellectual background and current theoretical dilemmas of marxism-feminism, setting the context for the main themes and contributions of the various chapters.

A.

The feminist challenge to marxism arose from complex roots. Analytically, it seized on the glaring fact that the subordination of women did not seem to be in any sense coterminous with capitalism. Not only had especially oppressive forms of patriarchy flourished in precapitalist societies; the

1

situation in many contemporary socialist societies indicated that the abolition of capitalism would not necessarily lead to patriarchy's demise.

Conventional Marxists had always approached the "Woman Question" as a function of women's position in the system of production. Early marxist analyses of women in capitalist society focussed on women's exclusion from wage work and their confinement to the "non-productive" sphere of housework as the key explanatory factors. The capitalist system, with its emphasis on private property and the commoditization of waged labor, was seen as the cause of this sexual division of labor. Engels, in *The Origin of the Family, Private Property and the State*, argues that women's subordination is a form of oppression resulting from the institution of class society and maintained because it serves the interests of capital. Male dominance, according to Engels, was inextricably bound up with capitalism and the ruling class, and would disappear with the advent of a socialist revolution and the wholesale entry of women into the waged labor market (Engels 1972).

In contrast, radical feminism, initially expressed by writers such as Shulamith Firestone (1971) and Kate Millett (1971), sought to understand women's oppression by focussing directly on sex: on male/female inequality in biological reproduction, conceived as a trans-historical fact, independent of and more important than class inequities. In Maureen Mackintosh's oft-quoted words, "the characteristic relation of human reproduction is patriarchy, that is, the control of women, especially of their sexuality and fertility, by men" (1977). But the danger in this approach was biological determinism, or the reduction of sex inequality to biology (Barrett 1980, p. 13). One answer seemed to be to distinguish sex as a biological category from gender as a social one. This insight can be traced to Gayle Rubin (1975), who proposed the term "sex/gender system" to denote the "set of arrangements by which the biological raw material of human sex and procreation is shaped by human, social intervention." Rubin interpreted the sex/gender system quite non-materially; it seemed, in her account, to be almost synonymous with the kinship system.

Two main issue areas were raised by feminists: patriarchy and reproduction. According to many, the former had its roots in the latter, rather than in production. A third feminist concern, one that did have to do with production and thus was of immediate interest to marxism, was unpaid "domestic labor" in the home, done largely by women. Christine Delphy, for example, located the origin of patriarchy in men's control over women's labor in the domestic arena: housework and childcare. Expanding Millett's argument that women's relation to the means of production differed from men's, since it is contingent upon marriage, Delphy gave the example of the divorced wife, the "displaced homemaker." Though such a woman may have been married to a man of the capitalist class, she did not usually own the means of production and was not therefore a full member of that class. Women's class position must be understood in terms of the institution of marriage, which is in fact a labor contract constituting a domestic mode of production and a patriarchal mode of exploitation (Delphy 1977; for critiques see Barrett and McIntosh 1979 and Molyneaux 1979).

Many feminist writers were strongly influenced by marxism; some considered themselves to be working within it. Yet feminism as a whole posed a profound challenge to marxism. In her influential essay, Heidi Hartmann (1981) argued against the subsumption of the feminist struggle into the supposedly "larger" struggle against capital. The "marriage" of marxism and feminism, if it were to continue, had to become an egalitarian one. The near-universal male dominance over women demanded explanation, and it was not sufficient to argue simply that it was functional for capital. Further, the relationship of women to men could not be adequately explained simply by reference to the relationship of women to the economic system. Categories of analysis were needed that were *not* "sex-blind."

Marxism struggled to respond to the feminist onslaught. The limitations of the concept of "patriarchy" were quickly noted: its universalistic character, its purported autonomy, its lack of division into types, variations or degrees, made it impossible to link to any changes in the organization of production or to any historical trends (Beechey 1979; Barrett 1980, pp. 10–19). Efforts were undertaken to attach this cultural and relational notion to a material base (Kuhn and Wolpe 1978). Roisin McDonough and Rachel Harrison, for example, agreed that patriarchy was "the form of control of the wife's labor and of her sexual fidelity," but they reasserted the subordination of this control system to economic class relations. At least under capitalism, the social relations of human reproduction were class-specific. True, the wife only "inhabited" her husband's class position, but "different contradictions inevitably arise for women inhabiting different class positions" (1978, pp. 34–37). This proposition seemed persuasive empirically, but it did not explore or resolve the fundamental question of the relation of reproductive to productive labor.

A still more traditional marxist view was reasserted in the work of Karen Sacks (1979), who quite explicitly saw reproductive relations as secondary. She argued that historical changes in the mode of production in Africa, and specifically in women's relations to the means of production as either "sisters" (equal members of a community of owners) or "wives," determined their wider political, personal and social status. Kin and marriage relations were seen as allocating material means, rather than children. Contrasting communal, kin corporate, and class or state societies, Sacks argued a well-constructed case against the notion of the universal subordination of women.

Most writers, however, saw a need to conceptualize women's status in reproduction as well as production. With reference to the contemporary

Third World this view was underlined in the writings of Lourdes Beneria and Gita Sen (1981; 1982). They called for a double focus on class and gender in studying the impact of development on Third World women, posing this approach in opposition to both neo-classical economic theory and cultural explanations of women's subordination such as those inherent in modernization theory. Beneria and Sen were less concerned to specify the exact interrelations between reproduction and capital accumulation than they were to point to aspects of women's position in the Third World, such as seclusion, bridewealth, dowry, and the "double day" of home and production work, which were unconvincingly explained in conventional marxist analysis.

Many of those committed to a merging of marxism and feminism found that the term "socialist feminism" best expressed their theoretical and political goals (e.g. Jagger 1983). Following Rubin and Hartmann, initial efforts at fusion were encapsulated in the notion of "dual systems," a patriarchal one, based in the household or family, in which women are controlled by men, and an economic one, in which workers are controlled by capitalists. Zillah Eisenstein and others attempted to specify the relation of the two spheres by arguing that patriarchy was functional for all modes of production: "Male supremacy, as a system of sexual hierarchy, supplies capitalism (and systems previous to it) with the necessary order and control" (Eisenstein, ed., 1979, pp. 27–28). Despite its insights, the dual systems approach underestimates the difficulty of achieving a synthesis of marxism and feminism. In some formulations the major analytic task is taken to be the relation between the "sex division of labor" and patriarchy in a cultural and social sense. In this version, the confusion remained over whether it was the sex division in production or in human reproduction which was most important. If it was the latter, then women's oppression was clearly quite separate from the sphere of production and as Iris Young objected, this dual system approach allowed traditional Marxism "to maintain its theory of production relations . . . in a basically unchanged form" (Young 1981, p. 49). In other formulations women's subordination was seen as related in some way to the mode of production and class structure, but little headway was made in the analysis of this question. Lise Vogel argued that "socialist feminists have worked with a conception of Marxism that is itself inadequate and largely economistic. At the same time, they have remained relatively unaware of recent developments in Marxist theory" (Vogel 1981, p. 197). As the alternative, Vogel offered the more traditional marxist view that human reproduction should be seen as one part of the reproduction of labor power, as part of the larger process of social reproduction (Vogel 1983).

Clearly, the theoretical status of biological reproduction had become central to the debate. Many feminists had turned to Engels in support of their contention that human reproduction was as important as production. Time and again this famous passage was quoted:

> According to the materialistic conception, the determining factor in history is, in the final instance, the production and reproduction of immediate life. This, again, is of a two fold character: on the one side, the production of the means of existence, . . . on the other side, the production of human beings themselves, the propagation of the species. The social organization under which the people of a particular historical epoch live is determined by both kinds of production . . . (1972, pp. 71–72)

Unlike much of Marx's writing, these lines seemed to give major recognition to human reproduction, anchoring it firmly as a material and social process; they avoided Marx and Engels' more usual characterization of the sex division of labor in childbearing and rearing as "natural." But they still did not solve the "dual systems" dilemma: what was the relationship between production and reproduction? What independent status, if any, did the system of human reproduction have?

Claude Meillassoux is perhaps the first modern marxist to give autonomous importance to human reproduction. The conditions under which capitalism arose, he wrote, did not point to the reproduction of human labor power as an urgent matter, either practically or theoretically. "Through the process of primitive accumulation it was solved straight off and the peasants migrant from European hinterlands contributed to rid the theoreticians of this extra worry" (1981, p. xii). But in precapitalist "domestic society," the somewhat vaguely defined "domestic community" in which reproduction takes place is not subordinate but is the very unit of production. The reproduction of labor is central. This is particularly so because the level of development of productive forces provides tools which are extensions of the human body rather than ones which subordinate it to non-human means of production. In domestic societies, and to a certain extent in all subsequent modes of production since they continue to depend on the domestic community for human reproduction, the social position of women derives from male control of their special reproductive powers. In the course of human evolution, the advanced domestic community succeeds in governing reproduction through the orderly circulation of women among men, that is, through patrilineality and patrilocality. Women are the actual means of reproduction. Rights to the progeny are granted to the husband's community; women cannot create descent relations or reproduce social ties: filiation only operates through men. Women are dispossessed of their children to the benefit of men; this exploitation of their reproductive capacities sets the stage for their second exploitation, their inability to acquire a status based on rights in the means of production.

One weakness of this account noted by marxist-feminists was that women seemed to be completely passive, apparently acquiescing in their own oppression (Mackintosh 1977). Another weakness, pointed out by Felicity Edholm, Olivia Harris and Kate Young (1977), was the failure to distinguish between the reproduction of human beings and the reproduction of the labor force, or those available to work. A common problem in all early marxist-feminist writings was the tendency to lump together three separate concepts: biological reproduction, reproduction of the labor force, and the larger process of reproduction of the social relations of production in the mode as a whole. Hindess and Hirst decried what they termed this "astonishing play on the word reproduction" (1976, p. 54). In rescuing discussion from this conceptual morass, Edholm, Harris and Young made a simple but critical point: in any mode of production the labor force, i.e. those available to perform socially productive labor, is socially constituted. Under capitalism, for example, it consists only of those who are forced to sell their labor power on the market. Human reproduction cannot be seen as directly equivalent to the constitution of the labor force. There are a set of intervening processes which categorize individuals as to their position in the labor process. Thus a central issue is the analysis over time of the relation between population levels (the outcome of society's, or men's, control over reproduction) and the organization of production (which determines the demand for labor).

Meillassoux seemed to be arguing that in all human societies subsequent to hunting bands and gynolocalities (excepting, in the future, advanced communism), men had *of necessity* controlled reproduction, and hence controlled women. Edholm, Harris and Young responded that even if some kind of social control over reproduction were a necessary feature of all societies, such control need not necessarily be by men, nor need it be inimical to the interests of women. Furthermore, control over reproduction was only part of the problem of controlling allocation to the labor force. Finally, the question of who controlled women's important work as material producers had been neglected. Reproduction had become an overloaded concept; to make it the all-inclusive explanation for men's control over women was unconvincing.

In addition to "patriarchy" and "reproduction," the marxist-feminists took up the question of "domestic labor" raised by radical feminists and this issue soon became a major preoccupation. Against early feminist efforts to depict housework under capitalism as a definite mode of production (Harrison 1973; Delphy 1977) producing only use values (Benston 1961), marxists were concerned to specify housework's contribution to capital, to demonstrate that it was functional to it. Wally Seccombe's major statement (1974) argued that the wage, which appeared to pay for the worker's labor only, was actually paying for all the labor that reproduced his labor power and that of his family. The wage has two parts, that which sustains the laborer, and that which sustains the housewife. Domestic labor *is* commodity production, in that it produces labor power, yet it is not subject to the law of value, it does not produce "surplus value," and therefore is not technically "productive." In reply, Jean Gardiner (1975) raised the question: if domestic labor did not contribute to surplus value, then how did capital stand either to gain or to lose by domestic labor? Further, she could see no intimation of exploitation in Seccombe's implied equal exchange between husband and wife. She asserted instead that the wife's labor time in effect adds to the wage time of the husband to yield the total subsistence or necessary level of labor time, and that domestic labor reduces the necessary part of the worker's labor time, or the value of labor power, to a level that is lower than the actual subsistence level of the working class. The difference is made up by the housewife's unpaid labor. Domestic labor thus contributes significantly to surplus value. This thesis has remained widely accepted. However, Susan Himmelweit and Simon Mohun (1977) have argued effectively that domestic labor, being private, non-capitalised labor as Harrison and Delphy point out, and not commodity production as Seccombe believes, cannot be subject to the capitalist law of value, and therefore any attempt to quantify or to compare domestic labor and wage labor, to ascertain how much or if at all it lowers the value of labor power, is simply impossible.

At first this convoluted debate may seem to have limited relevance to the position of women in the non-capitalist, non-western world, where the distinction between domestic and broader agricultural tasks is nearly impossible to make. "Domestic labor" as an analytic category was in effect created by capitalism, by the separation of most but not all social production from the household and kin group. To treat as domestic labor the superficially similar tasks performed by women in independent non-capitalist productive systems, in which little or no labor is valorised, is not only confusing, as Edholm, Harris and Young point out, it is fundamentally incorrect.

However, the domestic labor debate is relevant to the Third World because it raises the question not of an independent productive system, but of a subsidiary one. The economic relation of domestic labor to capital can be viewed as a particular case of the general problem of the articulation between dominant capitalist and subsidiary non-capitalist production systems. The constitution of capitalism as a world system has not yet meant the elimination of non-capitalist systems, it has more often meant their incorporation. Some have argued that domestic production within advanced capitalist societies may be a remnant, a survival of previously independent household production systems. If so, to dismiss a theory of articulation as impossible, as Himmelweit and Mohun do, seems to be abandoning efforts to understand a most important historical process.

For example: if women's domestic labor in households contributes to the reproduction of male labor power, if it reduces the value of male labor power enabling capital to pay wages below what they would be if domestic work were fully commoditized, then surely this effect will be intensified where women or households have some access to independent subsistence means. Thus, Carmen Diana Deere (1976) and others have argued that throughout the periphery of world capitalism, where men engage in some wage work but the household still has access to land, women's subsistence labor of all sorts contributes greatly to lowering the value of male labor power and increasing capitalists' profits. In contrast, in advanced capitalist societies the household's lack of access to subsistence means either that women must become wage earners too, or that male wages must rise to the level of the "family wage."

If the woman who does domestic work also enters the labor market, she is, Veronica Beechey (1977) argues, only "semi-proletarianised;" her situation is directly analogous to that of male migrant or semi-peasant proletarians so common on the periphery of world capitalism. Both are dependent on their own household production for part of the costs of their subsistence. Beechey pointed to three separate effects of such semi-proletarianized labor in the market: (1) its existence tended to lower the value of labor power in general; (2) such labor itself tended to have a lower average value than full-time labor since it lacked training and education; and (3) as Gardiner put it, this labor could be paid below its actual value since it was in part supported by the family. Such arguments are intuitively persuasive, despite the obstacles to making wage and household labor commensurate, and to pinpointing the relationship between value and wages.

The insight that the situation of the housewife in the developed capitalist nations is similar to that of the peasant and part-time worker in the Third and Fourth Worlds, proved a fertile one. Both, it is observed, engage in subsistence production for its use value, but at the same time some part of this use value must, because of the pervasiveness of capital, become commoditized. Like domestic labor, peasant production both is, and is not, capitalist, a paradox that has occasioned endless controversy within marxism. Veronika Bennholdt-Thomsen, for example, takes the view that extended reproduction, or capitalist accumulation, both determines and fundamentally depends on subsistence production (1981; but see also Smith 1984). Housewives and peasants together are the world's industrial reserve army, the "marginal mass" of capitalism, whose permanent absorption into wage labor cannot be foreseen in the near future.

Writers in the domestic labor debate usually included biological reproduction within the corpus of domestic work, under the label "generational reproduction." Doing so, however, introduced the two errors that Edholm, Harris and Young had so effectively identified: the conflation of human reproduction with the "reproduction of the mode of production," or with the reproduction of labor power. Bearing and rearing children who *might* become workers cannot be reduced to producing the commodity labor power.

In this Introduction, we propose an alternative conception: that children may be viewed as having both use value and exchange value under capitalist conditions. Either their labor power or their reproductive capacities might have exchange value. In any case, though, some part of the multifaceted "use values" of children must usually be marketed for the family and the individuals to survive. How much is marketed and at what rate is to a large extent determined by capital.

Considerations such as these suggest a new way of looking at human reproduction: as a distinct kind of labor and material production, which can be analyzed in ways analogous to a marxist analysis of the production of things. In the next section we explore this proposed new approach more fully, and suggest that a proper understanding and theoretical incorporation of human reproduction can result in a much-needed broadening and deepening of the marxist tradition.

B.

As a theoretical exercise, let us say that a mode of production is "an articulated combination of relations and forces of production structured by the dominance of the relations of production" (Hindess and Hirst 1975, p. 9). In any production process, Balibar tells us, humans are either laborers or non-laborers, and they are combined with nature, with the means of production, according to both a "property" connection and a "real appropriation" connection, which correspond respectively to the "relations of production" and the "forces of production" (Althusser and Balibar 1970, p. 215). The forces of production are a mode of appropriation of nature seen as a technical labor process, including various forms of division and cooperation between humans and between humans and tools or machines. Relations of production, on the other hand, depend on differential human ownership and control of means of production, which make possible differential human control of other humans. They result in a specific form of appropriation of surplus labor. All modes of production have both necessary labor, that sufficient to ensure the reproduction of the laborers, and surplus labor. In any mode of production the relations and forces of production form an articulated, intricately connected combination. It may be argued, however, that the relations of production are the logically primary element. In capitalism, for example, the class struggle may be the determinative factor, setting limits to the development of the forces of production.

How could the reproduction of new human beings be fitted into this basic neo-marxist schema? It should, as feminists have argued, be considered as in itself a particular form of production (cf. O'Brien 1981). In this productive process, the physical body of the woman above all, but to a certain extent that of the man, supported with food and shelter, constitutes the basic means of production. For women, the reproductive labor process consists of sexual intercourse, birthing, and lactation. For men it consists of sexual intercourse. The actual labor process can vary in technical complexity. On the one hand it might be termed "natural," if by that one means the process as it has emerged from centuries of primate evolution. In this case the body of the birth mother is the central instrument of production; the woman must be the primary laborer at least until the child is born. At this level of development of the forces of reproduction, there is a given imbalance in male and female tasks. As Meillassoux pointed out, one male may impregnate a large number of females without difficulty; the limit on productivity in a society is set by the number of women, not men.

Over time, however, humans have intervened in the reproductive process with increasingly sophisticated technologies: contraceptives, caesarian sections, *in vitro* fertilization. The technical forces of reproduction have profound implications for productivity and its social costs—for the number of live births per woman, and for infant and maternal mortality rates. They also have the potential to make obsolete the "naturally" given imbalance of male and female tasks in reproduction.

The relations of control, the "property connection" to the forces of reproduction, mediate the technical forces. All potential impacts of technical advances or technical imbalances depend on the relations of reproduction. From the point of view of the status of women, this is the crucial connection. The reproductive "labor force," who may and who must bear children, is socially controlled, often by men. The timing and extent of production may also be controlled by men. In sum, a biological mother, despite her given "real appropriation" connection with her child-bearing capacities, and despite her craft-like efforts to ensure a good product and reap its rewards, may find those capacities and/or their products effectively possessed and controlled by others. There is as yet no fully adequate explanation for this patriarchal state of affairs. One approach is that of Mary O'Brien, who argues that reproduction is a process of alienation for men, a separation from their "seed," and that therefore an effort to appropriate the child must be made. Paternity is most easily established through a coalition of men against women (O'Brien 1981, pp. 45–64).

By and large the regulation of reproduction has been through male-dominated social institutions of marriage and kinship. These enable the virtual non-laborers in the reproductive process, men, to appropriate for themselves a part of the "surplus" of reproduction, the benefits above and beyond those needed by the mother and child to survive. In high-fertility societies men may "accumulate" this surplus through large numbers of dependent women and children. The struggle over how many children to produce and who shall have rights to them is the content of the "class struggle" in the reproductive realm. However, the classes cannot simply be identified with the naturalistic biological categories of men/women. Class categories of controller/non-controller of the means of reproduction must and can be defined; the question of whether these are identical to the categories male/female is an empirical and historical one.

The approach to the conceptualization of reproduction which has been sketched so far differs from the sociological approach to sex differences. The distinction between the forces and the relations of reproduction is not equivalent to the sociological distinction between the terms "sex" and "gender," which refer to biological sex and the social/ideological construction of its meaning. Both the forces and the relations are constituted of ideational and material elements intertwined. Neither concept, forces of reproduction nor relations of reproduction, can be reduced to biological categories, unaffected by human mediation. Thus to attribute causal effectivity to the forces of reproduction is not biological reductionism, although the further one goes back in human history the more it may tend to become so empirically. But even in this case, the relations of reproduction would be preeminent, and they direct attention to the axis of control, a dimension which is conceptualised only indirectly in the sex/gender terminology. In the view presented here, the construction of the social meaning of masculinity and feminity at the level of ideology (gender) would depend on the social organization of reproduction and production.

The objective of feminism in general has been posed as the establishment of a theory of patriarchy, defined as the rule of men over women. The sociological approach has been to redefine the physiological categories into social ones; the problem then becomes the rule of the masculine gender over the feminine one. This strategy avoids biological reductionism by going to the other extreme of overemphasizing ideational constructs. In the materialist-based conception the feminist question has been whether in any or all societies there exist causal processes *specific to*, or categories of labor and resources *identical with*, the biological defined categories, such that these could be the cause of male control. The answer is still unclear, but the present analysis suggests categories of analysis that recognize the impact of human consciousness on the biology and technology of production, without denying that biology and technology may in themselves have certain pervasive effects. The social relations of reproduction clearly must have a complex, dialectical relation to the material facts of who does or does not do reproductive labor, and who does or does not reap its rewards.

This approach to the conceptualization of reproduction sets the stage for consideration of the problem of the relation between reproduction and production, but also indicates how complex such a question is. Does control over reproduction also produce control over who can or must enter the productive labor force? If so, it would seem to provide control over one of the most important factors in production, labor. Under capitalism, however, the size of the labor force would seem to be largely dictated by the organic composition of capital and by the business cycle. At another level, if the household is the effective unit within which reproductive decisions are made, what is its relation to the larger productive system? What determines whether household decisions are the basis of, or peripheral to, the larger economy? As to women, to what extent does a woman's pivotal role in reproduction set technical limits to her roles in hunting, cultivation or wage work, or does her common exclusion from these have to be explained as a consequence of the dominance relations in reproduction or production? These questions have only begun to be approached satisfactorily. Two recent efforts will illustrate the state of discussion in the field.

Wally Seccombe (1983) proposes three fundamental departments of production: (1) the production of the means of production; (2) the production of the means of subsistence; and (3) the production of labor power on a daily and generational basis. These articulate in a variety of ways within particular modes of production. Seccombe insists in the abstract that the forces/relations combination must be conceptualized for all three of these departments or productions, but then, surprisingly, he fails to do so. Instead he falls into the common marxist error of assuming that class, defined in relation to departments 1 and 2, is without question the prior analytic category, structuring all marriage, household formation and fertility patterns. He also assumes population levels represent directly the production of labor power. Accordingly, he moves directly to the analysis of "fertility regimes" for various laboring classes under capitalism, without specifying the theoretical status or effectivity of the forces/relations of reproduction. There is no mention of patriarchy here. Old-style demographic determinism, according to which population was an unexplained exogenous factor, is certainly unsatisfactory. But so too is the old-style marxist determinism of a class struggle conceived solely in terms of production. Population forces, Seccombe grants, may periodically come into conflict with "other elements" in a social formation and may spur technological changes leading to major transitions. But Seccombe actually gives us no compelling explanation why this should be so. In his conception, population levels are primarily determined by the economic need for labor.

At the other theoretical extreme, some feminists have argued for the concept of an independent "mode of human reproduction" (Bryceson 1980; Bryceson and Vuorela 1984). Bryceson and Vuorela write that "Just as the modes of production are constituted by distinct relations and forces of production, so too, [are] the modes of human reproduction . . . ," "labor power" in the one being similar to "sexual power" in the other. Their conception of the forces and relations of reproduction is similar to that proposed here, and is an advance over Seccombe's formulation. But it contains two critical problems. First, Bryceson and Vuorela give the mode of human reproduction (MOHR) a separate, autonomous status, outside the mode of production (MOP). The MOP/MOHR relation is one neither of subordination nor of dominance; rather they are separate entities, "each embodying a development dynamic unique to itself." Together they are referred to as a "mode of human existence." In contrast, we have proposed above that sexual power is not simply analogous to labor power, but rather *is* a form of labor power.

Moreover, a closer interrelation of the MOP and the MOHR seems inescapably necessary. Obviously, no material production can take place without human reproduction, and *vice versa*; each massively conditions the other. Theory must reflect this interdependence. The argument made by Banaji (1977) and others as to the need for any "mode" to be in principle autonomous and self-perpetuating is relevant in this context. A MOHR cannot in any way exist without a MOP; therefore it should not be called a "mode." The same of course is true of a MOP, although marxists have until now failed to recognise it. The most adequate concept would seem to be "mode of production/reproduction," containing within it the two different kinds of production.

Second, Bryceson and Vuorela consistently give primacy to the forces of reproduction rather than to the relations. They assert, for example, that the "demographic configuration" of the MOHR is determined by those forces, conceived as the degree of control of man/woman over nature. There is no perception that high fertility may be imposed by men where they gain disproportionate benefit from it, and may decline as it costs to men increase (for example, Folbre 1983). Bryceson and Vuorela argue, in fact, that the "material basis" of women's alienation is "the existence of children," and that "men are not the problem." They propose that where women bear fewer children, they are less alienated; this may be so, but only to the extent that men contribute little labor to the child production process, do not compensate women for the extra burdens and risks, and succeed in controlling the rewards. The social relations of reproduction are crucial. To deny this would be to say that women can never bear children without being oppressed. Or, it is to say that reproductive technology will inevitably liberate women, regardless of who controls such technology. Adequate feminist analysis must avoid technological as well as biological determinism.

Much work remains to be done in theorising the relations between production and reproduction, and in tracing their historical interconnections. As Joan Kelly (1983) has suggested, the stark oppositions so often posed by feminists between reproduction and production, between separation and participation, between sex and class, may in themselves be expressions of the nineteenth-century Victorian conception of divergent sociosexual spheres. Social conditions in the west have now changed; in most parts of the Third World they have always been different. Feminism may now be in a position to move forward with Kelly's "double vision," to articulate the systematic interconnectedness of women's lives.

C.

Many of the themes discussed above are played out in the papers in this volume. The opening contribution by Jeanne Henn approaches the issue of the grounding of patriarchy in material relations by proposing that patriarchy *is*, under certain historical circumstances, a mode of production. In the household production system she describes, gender relations *are* in fact class relations. This approach attempts to answer many of the criticisms feminists have made of Meillassoux's "domestic mode of production," Pierre-Philippe Rey's (1975) "lineage mode," and John Caldwell's (1982) "family mode." In Africa, the patriarchal mode has become integrated in subordinate fashion into world capitalism and articulates with it. Henn's approach differs, therefore, from the "world systems" view which sees household production/ reproduction strategies as ultimately defined only by capital, and which accords little or no role to gender/class struggles within the household.

Henn challenges the orthodox position of Himmelweit and Mohun that labor valuations in capitalist and non-capitalist modes are absolutely incommensurable. She argues that non-capitalist production is socially regulated and therefore value-creating; socially necessary or "abstract" labor can be defined for any mode, and so therefore can exploitation. Relative labor and consumption levels of individuals in households which participate in two modes can also be estimated.

In her empirical application of the patriarchal mode construct, Henn draws on her extensive field research on the southern Cameroon peasant economy and its sexual division of labor. But her model is couched in general enough terms to encompass a multitude of kin structures, descent systems and residence patterns; its potential applicabilty to a range of African and peasant societies is indicated. However, Henn leaves open the question of whether households more firmly subsumed under capital, such as those of wage and salary workers, can be analysed as modes of production, and their internal relations as class relations.

The patriarchal mode approach attempts to correct for the male-centered biases in earlier marxist analyses of precapitalist societies by making central exploitation within the household: between elder males, junior males, and women. It does this without resort to the naturalistic categories of male/ female; exploitation does not take place between the sexes precisely, but between social structural categories defined in terms of differential access to material means. Presumably, a biological woman might occupy the structural position of household patriarch (for example, African "female husbands") without disturbing the general validity of the argument. However, the link to biology, to the forces of reproduction, is the most problematic aspect of this approach. Biotechnical factors are not given much weight; forces of reproduction often seem to be reduced to relations, and reproduction to be subsumed under production. As in Meillassoux' work reproduction is said to be the central dynamic of the mode, but as with his model, the processes through which children become labor and incomes for the patriarch need further elaboration, as do the technical productive conditions which may press toward high fertility.

The paper by Nancy Folbre gives us another application of the patriarchal mode of production model: this time to the colonial and postcolonial transformation of Zimbabwe. This approach draws attention to aspects of the historical dynamic which have not been satisfactorily encompassed by the more usual capitalist/subsistence-peasant economies and proletarianization conceptions. It highlights two previously downplayed factors: the central role of women in subsistence and peasant production, and the impact of population growth. Folbre's aim is to interpret these as patriarchal social relations, and to place them "at the very center" of the explanation of the establishment, consolidation, and eventual breakdown of colonial control.

As she describes it, articulation took place between colonial capitalism and the indigenous "patriarchal-tributary" mode among the Shona and Ndebele. The patriarchal mode persisted and even flourished in the colonial Native Reserves; the interests of both capitalists and African patriarchs led to this outcome. But patriarchal relations and new health technologies together resulted in rapid population growth—a key factor in both the immiseration in the Reserves and the nationalist struggle. Patriarchal relations continue to pose obstacles to socialist reforms and economic growth. In the process of developing this historical account, Folbre makes several points about the patriarchal mode; for example, she emphasizes that the exploitation of women is measurable, and is derived from reproduction as well as production, but that on the other hand it tends to be socially limited by conceptions of proper kin relations.

As outlined above, one of the unsolved problems in the analysis of reproductive labor is its relation to the production of inanimate things. Jane Vock explores the answers to this question that are implicit in the

work of demographers Joel Gregory and Victor Piché, and John Caldwell.
She deftly exposes their work to the critical light of feminist analysis, in
which the subordination of women is grounded both in male dominated
relations of reproduction and in the technically sex-specific nature of child
bearing. For Gregory and Piché (1981) and Cordell and Gregory (1987),
fertility patterns in peripheral societies are a response to the capitalist
demand for labor; for Caldwell, they are a response to the persistence of
intergenerational wealth flows and, to the extent that they decline, to the
impact of Western ideas. Vock, in contrast, argues that fertility decision-
making within the household is an expression of differential gender relations.
She directs us to look at a more complex calculus of the costs and benefits
of children to women and to men, including labor contributions as well
as income flows. One might only add that to do such an analysis properly,
sophisticated economic tools will be needed. And, the distinction between
the use value of children in subsistence production and the potential value
of their labor power in capitalist production must be maintained.

Penelope Roberts includes the labor of children as part of the wider
problem of women's access to labor in West African farming systems.
Whereas the preceding three papers pose general theoretical issues, the
rest of the contributions analyze more political situations. Penelope Roberts
focuses more on women's position in West African agricultural production
compared to men's, with respect to access to a key resource, labor. She
thus raises the question of the social constitution of the labor force, pointed
to by Edholm, Harris and Young. In West Africa, rights in land are in
many cases more or less meaningless without rights in labor, and it is
precisely these rights that women lack in comparison to men. Why, Roberts
asks, do women manage "female farming" for household consumption, and
various "own-account" enterprises, with so little access to an important
means of accumulation, labor . . . either of children, other kin, slaves,
husbands or wage laborers? Low-intensity farming in which the wife individually does the bulk of the field labor should not, she reminds us, be
interpreted as "natural," either in the sense that it derives from the technical
requirements of ancient root crops or in the sense that women are prevented
by child-bearing from engaging in more intense labor. Rather, women,
unlike men, are constrained from mobilizing the labor of others' first by
their own labor obligations to their husbands, and second by the fact that
access to the labor of junior co-wives, work parties, or, in the past, slaves,
must be negotiated *through* the husband. The same considerations hold for
women's own-account enterprises. The heart of the matter is the "sexual
politics of labor" in the household. Roberts argues that the timing of
women's own-account enterprises over their life cycle is a consequence of
the constraints posed in early married years by labor obligations to husbands,
and that polygyny has been a means by which one generation of women

reallocates these obligations to junior wives in order to release more time for their own-account production. Even so, junior wives do not work *for* senior wives, but as their substitutes. Even in the past, when women were given access to domestic slaves, their use of them in production was contingent upon the husband's wishes. Differentiation among women was, and is, limited by the overarching gender relations of the household.

The spread of commodity relations in the African countryside has probably further reduced African women's access to labor. Commoditization of labor and land has enabled sons to enter wage labor or to establish their own enterprises, leading to greater household dependence on wives' labor. Schooling prepares children for success in the labor market, yet withdraws them from agricultural work. In this way, the growth of capitalism tends to intensify gender divisions and wives' subordination, "starkly revealing, as it were, wives' lack of ownership of the means of production yet their existence as unfree labor" (Roberts 1984, p. 178). Yet commodity relations may open up avenues through which wives may ultimately escape their existence as unfree labor.

Despite male opposition, in many areas a certain number of women were able to migrate to town. Whereas African elders and European colonial officials came to have a joint interest in keeping women's labor and reproductive capacities confined to agriculture, capitalists' interest depended on whether they wished to preserve male migrancy or to promote married, stabilized male labor. In colonial Zambia, most mining companies wished to do the latter; company housing and relatively high male wages promoted women's townward migration. The paper by Jane Parpart examines women's responses to this constellation of forces. She describes how the patriarchal re-imposition of marriage in the urban areas led to a strong counter-struggle by African women, waged in part through the urban courts. Parpart argues that women's interests lay in the reduction of marital constraints—in informal liasons, unregistered marriages, or marriages with a low bride-price. Personal liasons with men were of necessity women's main method of economic survival, since there were limited wage-earning and informal-sector opportunities. In addition, individual African men, freed from traditional constraints and not bound by the western ideology of monogamy, could not always be counted on to provide consistent financial support. Women benefited from mobility in the marriage market, which enabled them to get the best return for what domestic comforts they could provide. But in the end, both western and African norms of proper gender hierarchy were invoked in the successful effort to "stabilize" African town marriage. Women's independence was curtailed, but not entirely stopped. The struggle between the genders continued to challenge, at least to some extent, patriarchal dominance on the Copperbelt.

The essays by Parpart, Luise White and Janet MacGaffey provide a welcome counterweight to the tendency to see women simply as pawns in the patriarchal power game. While not denying the influence of structural constraints and systemic processes, they place more weight on the efforts of individual women to outwit the system, arguing that under certain conditions the efforts could be successful. Henn and Jacklyn Cock, on the other hand, tend to see individual actions as more limited by structural configurations.

Luise White describes the commodification of sexual and domestic labor which often took place in African urban contexts. Parpart, too, describes the usefulness to women of the extra-marital exchange of domestic services, although the temporary marriages she discusses clearly do not represent a true "free market" in such services. The essays by both White and MacGaffey describe how many Africans became petty entrepreneurs in the urban "informal sector," and in doing so found some measure of economic success and personal independence.

Luise White defines prostitution as domestic labor which has become a commodity sold by independent enterpreneurs. The labor of sexual intercourse is divorced from human reproduction, separated from kin and marriage relations; it is performed for pleasure and profit, frequently along with the provision of baths, hot meals and shelter, all adding up to the day-to-day re-production of male labor power. Prostitution in Africa has taken on expanded forms precisely because of the semi-proletarianization of male wage labor: the migrant labor system dictates that the generational reproduction of humans must take place hundreds of miles from the worksite, leaving the family unable to provide its usual daily services. In these conditions, the prostitute becomes the urban counterpart of the rural wife.

In the context of the many pejorative and morality-laden interpretations of prostitution, to recognise it as domestic labor is a considerable advance. But in relation to the "domestic labor" debate, it is confusing. While the tasks performed are the same as those done as "domestic labor" by the housewife under capitalism, the socioeconomic relations of production are clearly completely different. Prostitution is domestic labor commodified; it is unambiguously part of the bundle of commodities purchased and consumed by the laborer to reproduce his own labor power. Nevertheless, one can observe that it is women's biotechnically and historically given specialization in domestic tasks, whatever the variation in relations of production/ reproduction, that accounts for their continued identification with such work under petty commodity relations.

The core question in the domestic labor debate had been the relation of non-valorised household labor to the value of labor power, to capitalist surplus value. The labor of prostitutes, on the other hand, is valorised petty commodity production, and the relation of its price to the value of wages is more straightforward. This interrelation is an important theme of White's paper.

Like Luise White, Janet MacGaffey finds that the income-generating options that peripheral capitalism has offered to women in Zaire have been mostly in the informal sector. But in the case she examines, the opportunities have been notably profitable, owing to the transformation of the informal sector into a flourishing "second economy" under a weak postcolonial state. Whereas in the colonial era urban women subsisted and even profited from trade in food products, prostitution, and plot ownership, the enormous expansion of the second economy in the 1970's and 1980's provided even greater opportunities. Women have become accumulators, engaging in ivory smuggling, speculation, importing, and shipping foodstuffs such as beans, rice and fish from the interior. They invest in real estate, transport and rural plantations. Since in most cases no bank account, license or facility in French is required, a woman does not need her husband's permission, official approval, or much education to become successful; women can operate outside the usual mechanisms of male control. Many are unmarried; if married, their husbands often do not know the details of their business. Some are able to exchange sexual access for money or for political favors. Women make up approximately one-fourth of the larger business owners in Kisangani, and are predominant in market trade. MacGaffey argues that the successful businesswomen have risen from petty enterprise to become part of an emerging indigenous capitalist class.

The papers by White and MacGaffey raise the general theoretical question of the relations between petty commodity production and the capitalist mode of production proper. Both argue that women in the informal sector did well, some even becoming full-fledged capitalists. White suggests that prostitutes' earnings were always relatively high because of the prevalence of migrant labor, but that their earnings and work forms varied according to the level of real wages in the formal sector. This line of reasoning suggests that capitalism, at certain or perhaps all stages of its development, produces "places" for petty commodity production and benefits from its existence, even though at other times it acts to destroy it (for example, Gibbon and Neocosmos 1985). There are important controversies in the conceptualization of the petty commodity production/capitalist mode of production relationship. Both Henry Bernstein (1977) and William Roseberry (1978), for example, argue in relation to peasants that capital can exercise effective control of petty commodity production processes through the relations of exchange, for example through control of product prices, of terms of credit, and through competition with capitalist forms of production. Jairus Banaji (1977), on the other hand, has maintained that petty commodity production under capitalism is "value-producing," that is, that it has the same social function as wage labor and that the "price" which the producer receives is not a pure category of exchange but a "concealed wage."

The usual argument for the existence of a flourishing urban informal sector in Africa has been that it provides consumption goods to workers at a lower cost than capital could, thereby allowing wages to be kept low (for example, Davies 1979). It is also sometimes argued that it creates and maintains a "reserve army of labor," in this way also keeping wages down. However, to the extent that the informal sector and peasant commodity production offer good incomes as alternatives to wage labor, they may actually tend to raise wages by keeping workers out of formal employment and reducing competition. These arguments are analogous to those made with respect to domestic labor, and they seem as relevant to women's informal sector enterprises as to men's.

It is also often argued, following Marx, that capitalism will eventually destroy most petty commodity production. Independent artisans will be impoverished, and therefore forced into wage labor. Indeed, many seemingly independent workers in the informal sector are in fact dependent for materials, credit, tools and markets on large-scale enterprises (Scott 1979; Bromley and Gerry 1979). They have already become essentially out-workers, performing jobs more cheaply than wage workers could. The mechanisms of their subordination are many: falling commodity prices (Bernstein's "simple reproduction squeeze"); binding contractual arrangements; market competition from mass-production methods. However, it is difficult to imagine the informal sector in Africa being eliminated in the forseeable future; local food distribution and sexual services in particular, the two areas of women's greatest specialization, are the least likely to suffer competition from either international or local capitalists.

A few women in Africa have found jobs in the paid labor force. The papers by Sharon Stichter and Jacklyn Cock provide two views of wage earning women, middle class women in Kenya, on the one hand, and domestic workers in South Africa on the other. Both these papers also return to issues of analysis at the household level, initially raised for rural economies by Roberts and by Henn. All of these chapters assume that to understand the household, we must disaggregate it into its component individuals.

In Stichter's analysis of urban middle-class households, the most significant question has to do with the sexual politics of access to household income. She discusses the extent of income pooling among these largely dual-earner families, and suggests that the wife contributes a larger share of her (already smaller) wage income to the household. Forty-four percent of wives said they put almost all of their resources into the common fund. Clear patterns of differentiation of financial responsibility and of decision-making were apparent, husbands being responsible for the larger number of expenditures and having a larger role in household decision-making. As household income and women's incomes increased, however, more pooling of resources, more sharing of financial responsibility, and more jointness in decision-making was found.

The "sexual politics of labor" in the urban capitalist household refers only to housework, not to the agricultural or petty commodity production that Roberts described. Thus the "domestic labor debate" is directly applicable to African middle-class households. The bulk of household labor is still done by women; the husband takes only certain limited roles. The wife is permitted to hire a substitute, locally known as the housegirl, if she can pay for it herself. Only because of the access to such substitutes are wives able to combine marriage with the "liberation" of salaried employment. This strategy parallels the reallocation of labor to junior wives which Roberts describes for West African households.

Jacklyn Cock vividly describes the oppressive, coercive work relations which enmesh black women domestic servants in South Africa. Unlike the pattern in other industrializing societies, domestic work in South Africa does not serve to any great extent as a "bridging occupation" for women, a first step toward higher paying urban jobs. These workers are effectively "trapped"; influx control and high unemployment mean that the alternative is frequently starvation or grinding poverty in the homelands.

The institution of domestic service represents the incorporation of the wage relationship into the household. In this respect the work relations are quite different from those of other forms of wage labor; the worker, for example, is not exploited directly by a capitalist employer. In addition, as both Cock and Stichter point out, the employment of domestic servants does not disturb the sex division of labor in the household; in fact, it reinforces the exclusion of the male household head from domestic tasks, and leaves the wife still responsible for the administration of the work, the supervision of the servants. Ultimately, both the wife and the servant are subordinate to the husband. The domestic worker's contract is very personalised; she is often viewed as a "family dependent," though most black domestic workers in South Africa do not accept such a designation. The existence of a formal wage relationship, then, does not fundamentally change the character of the work as privatised household production.

Cock therefore correctly situates her analysis in the context of the domestic labor debate. She suggests such labor should not be labelled "unproductive" when considered in relation to its role in re-producing labor power. Black servants can be viewed as releasing white women's labor power, creating or perhaps transferring value which enters into the value of her employer's labor power. The work of black servants may also actually increase the value of white children's labor power, by contributing to their training and socialization. Of course, the work of domestic servants is performed for its use value to the employer as well, but the white family is forced to market a certain part of its human labor.

Cock makes the very important point that the relationship between domestic labor and capital must be recognised as historically variable. As pointed out in the first section above, the degree of contribution domestic work can make to the value of labor power will depend on the access the household has to independent means of production, for example land. It will also depend, as Cock's paper implies, on the level of development of capitalism. Where wages must be very low, either the contribution of domestic production must be greater, or more family members must be employed, or the necessary level of living must indeed be very low. As the working class struggles to increase the definition of what is necessary to their standard of living, and to raise wages, the wage may come to be high enough to cover a "non-working" wife, or the cost of a domestic servant. At still higher levels of wages, such things as commoditized child care and restaurant meals may be possible, that is to say, the fuller capitalization of domestic work. The two interpretations that Cock proposes for South Africa in this respect are not mutually exclusive. South African capitalism has made a concession to white workers in allowing them to build the cost of a black servant into their "necessary" means of subsistence. This has raised wage costs compared to a situation in which the unpaid white housewife did all the domestic work—although not enormously, given the appallingly low wages paid to African domestics. At the same time, the work of domestic servants in a sense subsidizes the white working class, since it lowers the cost of their subsistence in comparison to a situation of more fully capitalized domestic work. This concrete situation demonstrates the complexity of the relationship between domestic labor and capital, which varies according to the state of subsistence labor, of capital, and of the working class struggle to define its own necessary level of living.

As this brief review illustrates, the papers in this book are diverse, both in substantive focus and in the kinds of theoretical questions they raise. Topically, the essays are drawn from west, east, central and southern Africa, and touch on women's position in agricultural production (Henn, Folbre, Roberts), in petty commodity production (White, MacGaffey), and in wage labor (Stichter, Cock). They discuss the situation of women within marriage and the household, and outside it. One cultural/geographic area they do not cover is Islamic Africa. Methodologically, some of the essays employ primarily synchronic analysis (Henn, Stichter, Cock), while others are organised diachronically (Folbre, White, Parpart, MacGaffey); yet none takes the position that one kind of analysis precludes the other.

Theoretically, despite their particular foci, the essays share a common materialist approach to patriarchy, agreeing that it is grounded in reproduction and domestic labor, as well as productive labor. It is perhaps important to stress that none of our contributors writes as though patriarchy is timeless and unchanging. All of them see it as taking specific forms, bounded in space and time, but subject to change. Most see reproductive relations as changing in response to changes in productive relations, but Folbre and Cock are perhaps most explicit about seeing reproductive relations themselves as having an independent impact on the overall direction change.

As many of the papers in this book suggest, African women themselves have long recognized many of the difficulties that face them and have struggled to free themselves from both imperialist and patriarchal oppression. In recent years, African women have had to contend with allegations that feminism is really a new western onslaught on "African values," and is merely the latest aspect of western cultural imperialism. Given that western patriarchal conceptions were historically imposed on African societies and served in many ways to confirm or pervert traditional gender relations, the west cannot exempt itself from blame for the current situation of women in Africa. But western feminists too have struggled within their own societies against attempts to define feminism as a "bourgeois deviation," or as an aspect of ideology which will easily change as production structures are transformed (Roberts 1984). They have had to confront the political question of whether economic change or gender change should take priority. African women also face this double struggle, against economic exploitation on the one hand and patriarchy on the other, but the situation is complicated in that both conflicts have internal and external dimensions. African women themselves must and will decide which of their needs is to be given priority at any given moment, and whether external struggles should take precedence over internal ones. Western feminists can only strive to understand and support their efforts.

Notes

1. The editors recognize the wider tradition of socialist-feminism, but choose the term marxist-feminism deliberately, to indicate that we see ourselves working to a large extent within the specifically marxist tradition, and arguing for the possibility of marxism's internal transformation.

References

Althusser, Louis, and E. Balibar. (1970) *Reading Capital*. London: New Left Books.
Banaji, Jairus. (1977) "Modes of production in a materialist conception of history." *Capital and Class*, 3: 1–44.
Barrett, Michele. (1980) *Women's Oppression Today*. London: New Left Books.
Barrett, M. and Mary McIntosh. (1979) "Christine Delphy: Towards a Materialist Feminism?" *Feminist Review*, 1.

Beechey, Veronica. (1977) "Some Notes on Female Wage Labour in the Capitalist Mode of Production," *Capital and Class*, 3.

————. (1970) "On Patriarchy," *Feminist Review*, 3

Beneria, Lourdes. (1979) "Reproduction, Production and the Sexual Division of Labour," *Cambridge Journal of Economics*, 3, 3 (Sept): 203–25.

Beneria, Lourdes, and Gita Sen. (1981) "Accumulation, Reproduction and Women's Role in Economic Development: Boserup Revisited," *Signs*, 7, 2: 279–298.

————. (1982) "Class and Gender Inequalities and Women's Role in Economic Development: Theoretical and Practical Implications," *Feminist Studies* 8, 1 (Spring): 157–176.

Bennholdt-Thomsen, Veronika. (1981) "Subsistence Production and Extended Reproduction," pp. 16–30 in K. Young, C. Wolkowitz, and R. McCullagh, eds., *Of Marriage and the Market: Women's Subordination in International Perspective*. London: CSE Books.

Benston, Margaret. (1961) "The Political Economy of Women's Liberation," *Monthly Review*, 21, 4: 13–27.

Bernstein, Henry. (1977) "Notes on Capital and Peasantry," *Review of African Political Economy*, 10 (Sept-Dec.).

Bromley, Ray and C. Gerry. (1979) *Casual Work and Poverty in Third World Cities*. New York: Wiley and Sons.

Bryceson, Deborah. (1980) "The Proletarianization of Women in Tanzania," *Review of African Political Economy*, 17 (January-April): 4–27.

Bryceson, Deborah and Ulla Vuorela. (1984) "Outside the Domestic Labor Debate: Towards a Theory of Modes of Human Reproduction," *Review of Radical Political Economics*, 16, 2 and 3: 137–166.

Caldwell, John. (1982) *Theory of Fertility Decline*. New York: Academic Press.

Davies, Rob. (1979) "Informal sector or subordinate mode of production? A model," pp. 87–104 in Bromley and Gerry, eds., *Casual Work*.

Deere, Carmen Diana. (1976) "Rural Women's Subsistence Production in the Periphery," *Review of Radical Political Economics*, 8, 1: 9–17.

Delphy, Christine. (1977) *The Main Enemy* London: Women's Research and Resource Center.

Edholm, Felicity, Olivia Harris and Kate Young. (1977) "Conceptualising Women," *Critique of Anthropology*, 3, 9 and 10: 101–130.

Eisenstein, Zillah, ed. (1979) *Capitalist Patriarchy and the Case for Socialist Feminism*. New York and London: Monthly Review Press.

Engels, Frederick. (1972) *The Origin of the Family, Private Property and the State*. E. Leacock, ed. New York: International Publishers.

Firestone, Shulamith. (1971) *The Dialectic of Sex*. New York: Bantam Books.

Gardiner, Jean. (1975) "Women's Domestic Labour," *New Left Review* 89: 47–58.

Gibbon, Peter and Neocosmos, Michael. (1985) "Some Problems in the Political Economy of African Socialism," pp. 153–206 in H. Bernstein and B. Campbell, eds., *Contradictions of Accumulation in Africa*. Beverly Hills: Sage.

Gregory, Joel and Victor Piché. (1981) "The Demographic Process of Peripheral Capitalism Illustrated with African Examples," Montreal: McGill University Center for Developing Area Studies.

Harrison, John. (1973) "The Political Economy of Housework," *Bulletin of the Conference of Socialist Economists*. London.

Hartmann, Heidi. (1981) "The Unhappy Marriage of Marxism and Feminism: Towards a More Progressive Union," pp. 1–41 in L. Sargent, ed., *Women and Revolution: A Discussion of the Unhappy Marriage of Marxism and Feminism*. Boston: South End Press.

Himmelweit, Susan and Simon Mohum. (1977) "Domestic labour and capital," *Cambridge Journal of Economics*, 1: 15–31.

Hindess, Barry, and Paul Hirst. (1975) *Pre-Capitalist Modes of Production*. London: Routledge and Kegan Paul.

———. (1977) "A Reply to John Taylor," *Critique of Anthropology*, 8:

Jaggar, Alison M. (1983) *Feminist Politics and Human Nature*. Totowa, N.J.: Rowman and Allanheld.

Kelly, Joan. (1983) "The Doubled Vision of Feminist Theory," pp. 259–270 in Judith L. Newton, Mary P. Ryan, and Judith Walkowitz, eds., *Sex and Class in Women's History*. London: Routledge and Kegan Paul. (First published in *Feminist Studies*, 5, 1 (Spring).

Kuhn, Annette and AnnMarie Wolpe, (eds.). (1978) *Feminism and Materialism*. London: Routledge and Kegan Paul.

Mackintosh, Maureen. (1977) "Reproduction and Patriarchy: A Critique of Meillassoux, 'Femmes, Greniers, et Capitaux,'" *Capital and Class*, 2: 119–127.

McDonough, Roisin and Rachel Harrison. (1978) "Patriarchy and Relations of Production," pp. 11–41 in Kuhn and Wolpe, eds., *Feminism and Materialism*.

Meillassoux, Claude. (1981) *Maidens, Meal, and Money: Capitalism and the Domestic Community*. Cambridge: Cambridge University Press.

Millett, Kate. (1977) *Sexual Politics*. New York: Avon Books.

Molyneaux, Maxine. (1979) "Beyond the Domestic Labour Debate," *New Left Review*, 116 (July-August): 3–27.

O'Brien, Mary. (1981) *The Politics of Reproduction*. London: Routledge and Kegan Paul.

Rey, Pierre-Philippe. (1975) "The Lineage Mode of Production," *Critique of Anthropology*, 3 (Spring): 27–79.

Roberts, Penelope. (1984) "Feminism In Africa: Feminism And Africa." *Review of African Political Economy*, No. 27/28: 175–184.

Roseberry, William. (1978) "Peasants as Proletarians." *Critique of Anthropology*, 11, 3 (Spring): 3–8.

Rubin, Gayle. (1975) "The Traffic in Women: Notes on the 'Political Economy' of Sex" in R. Reiter, ed., *Toward an Anthropology of Women*. New York: Monthly Review.

Sacks, Karen. (1979) *Sisters and Wives*. Urbana: University of Illinois Press.

Sargent, Lydia. (ed.). (1981) *Women and Revolution*. Boston: South End Press.

Scott, Alison. (1979) "Who are the self-employed?" pp. 105–132 in Bromley and Gerry, *Casual Work*.

Seccombe, Wally. (1974) "The Housewife and her Labour under Capitalism," *New Left Review* 83 (Jan.-Feb.): 3–24.

———. (1983) "Marxism and Demography," *New Left Review* 137 (Jan.-Feb.): 22–47.

Smith, Joan. (1984) "Non-Wage Labor and Subsistence," pp. 64–89 in J. Smith, I. Wallerstein and H. Evers, eds., *Households and the World Economy.* Beverly Hills: Sage.

Stichter, Sharon. (1985) *Migrant Laborers.* Cambridge: Cambridge University Press.

Vogel, Lise. (1981) "Marxism and Feminism: Unhappy Marriage, Trial Separation or Something Else?" pp. 195–218 in Lydia Sargent, ed., *Women and Revolution.* Boston: South End Press.

――――. (1983) *Marxism and the Oppression of Women.* New Brunswick, New Jersey: Rutgers University Press.

Young, Iris. (1981) "Beyond the Unhappy Marriage: A Critique of the Dual Systems Theory," pp. 43–70 in Lydia Sargent, ed., *Women and Revolution.* Boston: South End Press.

2

The Material Basis of Sexism: A Mode of Production Analysis

Jeanne Koopman Henn

Most social scientists draw a sharp distinction between class and gender. Marxists, for example, identify classes by analyzing the social relations of production, while marxist feminists analyze gender in the context of the relations of human reproduction. This essay argues that class and gender are more intimately related than these formulations suggest and that we can better understand both by analyzing them in the same theoretical framework.

A major problem with the marxist feminist approach to gender analysis is the fact that different analysts define the "relations of human reproduction" differently. Some emphasize childbearing and rearing (Bryceson and Vuorela 1984), others stress the social as well as the daily and generational reproduction of the labor force (Beneria 1979), and many focus directly on male dominance, defining the relations of human reproduction as "the methods by and degrees to which men retain control over individual women as unpaid domestic labor, sexual 'partners' and bearers and rearers of children" (Roberts 1984, p. 179). While most feminists agree that men dominate the relations of human reproduction, they do not necessarily agree on why this should be so. As Roberts points out, "it is still a matter of much debate within feminist analysis whether . . . male control of the relations of human reproduction is to be explained in terms of men's material interests or lies purely in the realm of ideology" (Roberts 1984).

This essay develops a theoretical approach to the analysis of class and gender which demonstrates that under certain conditions, gender relations actually become class relations. I employ a marxian mode of production analysis (developed here in a manner emphasizing the interdependence of economic, political and ideological factors in the constitution and reproduction of class relations) to establish the fundamental class nature of a form of gender relations which result in women's subordination and exploitation.

Part two of the essay develops the concept of a patriarchal mode of production. The patriarchal mode is a theoretical model of class relations between a class of patriarchs who, as heads of households, control the access of other household members to the means of production and a class of patriarchal dependents, wives and working children, who gain access to the means of production and consumption by providing surplus labor to the class of patriarchs. This model should not be regarded as an assertion that all types of gender relations can be analyzed as class relations. They cannot. The purpose of developing a concept of a patriarchal mode of production is to provide explicit theoretical criteria which allows us to determine when and how male control of the relations of household production and of human reproduction become class relations. My basic argument is that an articulation of modes of production framework, which includes the patriarchal mode of production, can produce a more rigorous and comprehensive understanding of the socio-economic effects of male dominated gender relations than can be gained by supplementing class analysis with a theoretically different form of gender analysis.

I first felt the need for a concept which could account for class relations within household family production units while working out an articulation of modes of production model for an analysis of the rural economy of southern Cameroon.[1] In searching for a model to guide an analysis of the social relations of production in the stateless, precolonial societies of this area, I found that mode of production concepts such as the domestic mode of Meillassoux (1975), the primitive communist mode elaborated by Hindess and Hirst (1975) and even the lineage mode of P.P. Rey (1979) did not capture the classlike relations between Beti patriarchs, on the one hand, and their wives and male dependents, on the other. In the precolonial Beti case, patriarchal control over economic resources was associated with heavy labor for women, some work for male dependents, and little or no work for patriarchs. The classless domestic and primitive communist modes cannot explain economic exploitation associated with lack of free access to the means of production, while the lineage mode tends to ignore the systematic subordination and exploitation of women by identifying the primary class contradiction as that between patriarchs and male cadets. These models are therefore of limited value to an analysis of social relations of production characterized by the subordination and economic exploitation of women.

This essay develops a concept of the patriarchal mode of production and discusses its relevance to the analysis of patriarchal stateless societies in precolonial Africa. It argues that the theory of the patriarchal mode of production can contribute to a clearer understanding of interrelations between patriarchal classes and the classes of other modes of production.

I advocate working with an articulation of modes of production framework in order to focus analysis on the processes of class formation, alliance, and transformation over time and on the related processes of the transfer of surplus labor and the accumulation of economic resources which are the results of these processes.

A major theme in the articulation of modes of production literature has been the analysis of the effects of the process of capitalist expansion on noncapitalist modes.[2] A minor, even neglected, theme has been the analysis of the effects of the reproduction of noncapitalist modes on capitalism. Much of the work on the effects of noncapitalist relations, gender relations in particular, has been conducted outside the modes of production framework. Marxists and marxist feminists have attempted to assess the effects of domestic labor on the process of wage and profit formation in the well known domestic labor debate, which analyzes how labor performed by women in the home can be related to labor and value generated in a capitalist enterprise. The domestic labor debate came to an impasse in the late 1970s when most participants accepted the orthodox marxist assertion that non-commodity producing labor (housework, childcare, subsistence agriculture, etc.) is incommensurable with capitalist wage labor.[3] This position not only stymies a Marxist analysis of the interrelations between domestic labor and capitalism, it also precludes *any* analysis of surplus labor transfer between capitalist and non-commodity producing modes. I address this fundamental theoretical problem in the next section.

Other problems in the mode of production literature result from differing formulations of the fundamental concepts of mode of production theory and consequent inconsistencies in the characterization of particular non-capitalist modes. Because the concept of a mode of production has been elaborated in a variety of ways, in section one of the essay I define the general theoretical structure of mode of production analysis that is used in developing the patriarchal mode in section two. The concepts of socially necessary labor and surplus labor are discussed in some detail and the incommensurability of commodity producing labor and subsistence labor (production of goods and services for own or family use) is challenged.

The second section constructs a model of the patriarchal mode of production. The section begins, however, with a critique of recent attempts to apply a concept of a primitive communist or communal mode of production to the analysis of precapitalist societies where the conditions of access to the means of production are systematically different for male heads of households, women and other household dependents. It is argued that the model of the patriarchal mode of production is a far better guide to empirical research in these situations than the classless primitive communist mode.

Section three draws on African case materials to demonstrate the relevance of the patriarchal mode to an historical materialist analysis of many precolonial, colonial, and contemporary African societies. While I do not argue that the patriarchal mode is relevant to the analysis of types of household-based production in Africa, I do suggest that it has wider relevance than current formulations of classless, communal modes.[4]

Mode of Production: General Concepts

Mode of production theory focuses on the material, i.e. economic, basis of oppression and exploitation and on the class nature of social struggle and change. The version of mode of production theory adopted here has the following major analytical goals:

1. identification of class relations and their associated flows of surplus labor.
2. identification of the economic, political, and ideological sources of class power, class interests, and class conflict.

Mode of production theory *identifies* class relations in the sphere of the economy but also asserts that class relations cannot be established or continually reproduced without the social reproduction of corresponding political and ideological relations.

The Social Relations of Production

The social relations of production of a particular mode of production describe the social distribution of control over the means of production and the associated transfers of surplus labor between classes. The marxian definition of class posits an exploitative relation between the possessors and the non-possessors of a society's economic resources or means of production (Marx 1967, p. 791; Balibar 1970). By obtaining effective possession of at least some of the means of production essential to the reproduction of the economy, a dominant class is able to expropriate surplus labor from a subordinate class which is deprived of direct access to the means of production.

Two concepts are crucial here: the notion of effective possession of the means of production and the concept of surplus labor. Effective possession has been well defined by Hindess and Hirst (1977a, p. 65):

> Classes depend upon a relation of effective possession of the means of production and distribution. Effective possession is the capacity to control the functioning of these means in the process of production and the capacity to exclude others from their use.

In general it is only certain of the means of production which are exclusively possessed and are denied to others to whom they are necessary conditions of production, denied except on the terms of an economic relation of payment for their use.

The *economic relation of payment* refers to the ability of the effective possessors of the means of production to expropriate surplus labor from the non-possessor class. In a famous passage from the third volume of *Capital* (1967, p. 791) Marx emphasized the centrality of the concept of surplus labor to the analysis of class relations:

The specific economic form in which unpaid surplus labor is pumped out of direct producers determines the relationship of rulers and ruled, as it grows directly out of production itself and, in turn, reacts upon it as a determining element.

But just what is surplus labor? This is a question too often eluded, perhaps because it raises all the problems of defining labor and labor values which plague the vast literature on value theory.[5] A definition of surplus labor is both essential (because it is the *raison d'etre* of class relations) and complex (because it requires a prior definition of socially necessary labor). I will return to the problem of defining surplus labor after reviewing other central concepts of mode of production analysis: the means and forces of production, the role of ideological and political factors in the constitution and reproduction of class relations, and the concepts of articulation and dominance among theoretically distinct modes.

At any one point in time, the classes involved in a particular mode of production make use of a specific set of economic resources, such as land, machinery and labor (the means of production) in a labor process which embodies a certain level of technology and organization (the forces of production). The economic resources, the technology with which they are used, and the organization of the labor process may vary from one manifestation of the mode to another. It is the social relations of production, the form of possession of the means of production and the manner in which surplus labor is appropriated, which identifies a mode of production and its constituent class relations.[6]

When a dominant class can continually reproduce its social control over the means of production, it is able to set the terms upon which the subordinate classes gain access to the means of production. Even though these terms of access may appear to be accepted voluntarily by the subordinate class, if an exploitative class relationship exists further analysis will reveal that the subordinate class is politically or ideologically denied alternative forms of access which would involve less performance of surplus labor.

One of the more serious problems in the mode of production literature has been a relative neglect of the ideological and political aspects of the social relations of production. Outside the modes of production school, some marxist theorists have emphasized the fundamental importance of political and ideological relations to the constitution and reproduction of class relations (Bowles and Gintis 1982; Laclau 1977; Katz 1980). These insights need to be fully incorporated into mode of production analysis because the social relations of production, based as they are on a particular social distribution of control over economic resources, cannot be adequately defined without reference to the political and ideological relations which make exclusive control over certain means of production and distribution possible. No exploiting class can wield the kind of power which enables it to extract surplus labor unless that power is backed by political institutions which reproduce class dominance and by ideological, cultural, or psychological practices which legitimate it.[7]

Articulation of Modes of Production

The question of how coexisting modes of production affect one another has only recently received sustained theoretical and empirical attention (Wolpe 1980; Hindess and Hirst 1977a; Katz 1980). The developing theory of articulation of modes of production attempts to identify the points of interaction among modes and the effects of interaction on the reproduction of distinct modes. In some cases articulation may involve only market relations among classes formed in different modes; in others it may be much more intimate and complex, as when the same person or members of the same household are involved in two or three modes of production (Marx 1967, pp. 129–152). The existence of multiple class identities seriously complicates an empirical analysis of class interests. If the same persons or population groups have multiple and possibly contradictory class interests, possibilities for united class action may be seriously reduced. Such problems have been noted, for example, within the capitalist working class when male workers have opposed the interests of female workers (Hartmann 1979). An articulation of modes of production model can stimulate a more insightful analysis of this type of problem.

Some critics of the notion that several modes of production can coexist in a single society argue that the concept of articulation implies that all modes in a social formation must reproduce themselves autonomously (Banaji 1977; Bernstein 1979). These critics imply that to exist all modes of production must be capable of reproducing not only their material means of production and labor force, but also the political and ideological practices necessary to the social reproduction of their distinct social relations of production. This view assumes that to define a theoretical mode as a guide to empirical analysis, one must be able to show empirically that articulation with other modes has no effect on the autonomous reproduction of that mode. The major concern of these writers seems to be that the concept of articulation fails to recognize how historical examples of non-capitalist modes have been penetrated, transformed, and subordinated, if not fully destroyed, by capitalism.

In my view, the development of rigorous concepts of non-capitalist modes of production in no way impedes an analysis of how their historical counterparts have been fundamentally altered as a result of interaction with the capitalist mode. The historical process of articulation may even have obliterated the reproductive autonomy of a subordinate mode, but this does not mean that its characteristic social relations of production and their political and ideological requisites do not survive and continue to affect the development of the dominant mode. To properly analyze and assess the continuing impact of subordinate modes, we need appropriate and well developed concepts of such non-capitalist modes.

Empirically, when two or more modes of production coexist at a particular moment in time, one is likely to dominate. A mode can be considered dominant when its ruling class is able to compel the performance of surplus labor by classes of the subordinate modes and can dispose of the product of that surplus labor according to its own priorities. The theory of articulation is particularly important to the conceptualization and empirical analysis of the transfer of surplus labor among modes of production. To understand the development of a dominant mode and the impoverishment or decline of a subordinate mode, we need to be able to define and in some sense quantify the performance and appropriation of surplus labor among the classes of the interacting modes. An analysis which identifies and measures (however roughly) the material advantage various population groups obtain from the reproduction of class power in subordinate modes and from class alliances with the dominant class of another mode helps to clarify the material bases of political and ideological class behavior. We must return, then, to the problem of defining and measuring surplus labor.

Surplus, Necessary, and Socially Necessary Labor

To define surplus labor, Marxists have traditionally begun by dividing total labor time into necessary labor time and surplus labor time. In most cases necessary labor is defined first, and surplus labor is derived as a residual. Resnick and Wolff (1982, p. 2), for example, define necessary labor as the "time measured expenditure of human brain and muscle required to produce the performers of surplus labor." Surplus labor is then defined as that portion of total labor performed which is above and beyond necessary labor time. (See also Hindress and Hirst 1975, p. 26).

The residual definition of surplus labor poses serious theoretical and empirical difficulties for the analysis of situations in which individuals participate in more than a single mode of production. If the laborers in one mode, such as the capitalist, are reproduced (on a daily as well as a generational basis) only *in part* by consuming commodities purchased with the wage, their necessary labor cannot be estimated by the exchange value of their wage basket. But even if we attempt to assess all the components of a laboring class's consumption, when individuals or groups participate in multiple modes of production, we have no way to determine how much of their overall necessary labor is performed in each interacting mode.

It is both less ambiguous and analytically more powerful to define necessary labor as the labor hours embodied in the products and services the laborer receives for his or her work in a particular mode and to define surplus labor as the labor time which the class with effective possession of the means of production expropriates from the laboring class. This approach makes it absolutely clear that the relationship between necessary and surplus labor depends on the state of the class struggle. It also allows a precise definition of the rate of exploitation (the ratio of surplus labor to necessary labor) for each mode of production. It does not, however, permit the estimation of a single rate of exploitation for actual groups of people (male or female peasants, wage workers, or petty commodity producers) who participate in two or more modes of production.

A comprehensive analysis of the articulation of modes in an actual society requires a means of assessing the relative impact of total surplus labor expropriation on the laboring and the resource controlling groups of that society. In situations where commodity producing modes, such as the capitalist, interact with subsistence (own use) and commodity producing modes, such as the patriarchal or feudal, this task becomes intractable only if Marx's designation of commodity producing labor as "abstract" and labor for own use as "concrete" is interpreted as precluding their commensurability.

The Commensurability Problem

Those who assert the incommensurability of subsistence and commodity producing labor (e.g. Himmelweit and Mohun 1977) base their case on Marx's distinction, in the first chapter of *Capital*, between use-value (the utility of a product) and exchange-value (the quantity of abstract labor embodied in a product). The incommensurability argument suggests that the "concrete" labor time embodied in subsistence products is merely what Marx called individual labor, i.e., work done to satisfy personally determined wants. Since individual labor is personally, rather than socially determined, it cannot produce exchange-value. The exchange-value of a product, as Marx clearly argued, does not depend on the precise amount of time its individual producer(s) took to make it, but rather on the amount of "socially necessary labor" embodied in it. Socially necessary labor is defined as the amount of labor "required to produce an article under the normal conditions of production and with the average degree of skill and intensity prevalent at the time" (Marx 1906, p. 46).

While all Marxists agree that individual labor times are not the determinants of value, I still argue that we have no basis in Marx for asserting that abstract, commodity producing labor is the only form of socially necessary labor. Just as an individual's commodity producing labor time may either exceed or fall short of the socially necessary, abstract labor he or she is performing, so subsistence labor (labor whose product is directly consumed rather than marketed) should be understood as having a socially necessary component determined, as Marx suggested, by the "normal conditions of production prevalent at the time" and an individual component which varies with the skill and motivation of the individual laborer. Thus, subsistence labor is not merely individual labor, i.e. labor which is personally determined, nor are subsistence products devoid of socially necessary labor.

The argument that subsistence labor is not commensurable with abstract, commodity producing labor is often based on the claim that there is no means of socially regulating nonmarket-oriented labor time, i.e., no social mechanism to force subsistence producers to use the most efficient production methods available in the way that market competition forces commodity producers to be efficient (Himmelweit and Mohun 1977). It is therefore concluded that the amount of socially necessary labor time embodied in subsistence products cannot be measured. I reject this argument.

As Bryceson (1983) argues in an important theoretical elaboration of the concepts of use-value and exchange-value, subsistence production to meet family consumption and service needs *is* socially regulated. Even though the amount of time spent on tasks such as growing food for family consumption, cooking and childrearing varies across individuals and over time, the social norms regulating household subsistence labor can be defined for any particular historical period. These norms are expressed in social definitions of gender and age specific subsistence labor responsibilities and enforced by socially defined sanctions for disciplining individuals who fail to meet them. Socially shared and socially reproduced practices regulating behavior in the domestic sphere determine the normal conditions of household subsistence production just as surely as market competition determines the normal conditions of commodity production.

If we can agree that the social rationalization of labor time is not unique to commodity production, there is no basis for the capital centric assertion that only commodity producing, abstract labor creates value. As Bryceson has argued, " 'concrete' labor creating use-values is as social, objective and value-creating in nature as 'abstract' labour, the only difference being that it is mediated by social institutions other than the market" (Bryceson 1983, p. 37). Marx made the distinction between use-value and exchange-value in order to define the nature of the capitalist mode. It was hardly his intention to preclude an analysis of the articulation between capitalist and non-capitalist modes or an analysis of the impact of that articulation on the rate of exploitation.

Commensurability between commodity producing (abstract) and subsistance (concrete) socially necessary labor makes it possible to analyze how the articulation of modes of production affects the labor and consumption of different categories of family members in households where different modes interact. When subsistence and commodity producing labor can be compared, exchanges of goods and services among family members can be quantified and rates of exploitation can be estimated not only for specific modes of production but also for the actual people whose lives are structured by multiple sets of class relations (See Folbre 1982, for a further development of this approach.)

The Patriarchal Mode of Production

The Communal Versus the Patriarchal Mode

The social relations of household based production are commonly analyzed with an implicit or explicit model of a communal or a primitive communist mode of production. In these models, the unit of production may be a single household or a group of related households, but whatever its size, all persons born into the kinship structured production unit are assumed to have "an equal share and rights in the resources of the group" (Macfarlane 1978, p. 106). The assumption of communal possession of the means of production implies that there is no basis for denying any family member free access to the means of production; thus, there is no material basis for intrahousehold exploitation and no basis for class. A key empirical question in applying this model is the extent to which the assumption of communal access to the means of production is valid.

While I do not question the legitimacy of a theoretical concept of a communal mode of production, I am disturbed by the extent to which substantial inequalities in labor times and consumption among different categories of adult household members are considered compatible with communal possession of the means of production. Hindiss and Hirst's theoretical discussion of the primitive communist mode, for example, goes as far as to deny that differences in power, consumption or labor times have any importance in the identification of class relations (Hindess and Hirst 1975, pp. 21–78). The "complex redistribution variant" of their model of primitive communism identifies a group of elders whose functions are the coordination of labor and the regulation of marriages. Hindess and Hirst suggest that these functions can "involve the existence of forms of coercion" which may even include the sale of dependents into slavery, but elders are nevertheless said *not* to have the power of a dominant class.

When challenged (e.g. by Taylor 1975), Hindess and Hirst (1977b) consistently refuse to consider regulatory and coercive "functions" as means by which elders limit the access of male and female dependents to the means of production. Why? Perhaps because these theorists feel compelled to defend their definition of the social relations of production of primitive communism which "*requires* that property be vested in the productive community" (1975, p. 50 emphasis added). The logic of the model does not allow *any* behavior on the part of elders to be considered as a means of restricting access of other community members to the means of production.

If the communal mode of production is to be legitimately employed in the analysis of family, household, or lineage-based production, communal possession of the means of production must be demonstrated rather than merely asserted. One would expect communal possession to denote situations in which all adult members of a community participate in decisions concerning the use of the means of production. To assert communal possession in the face of coercion or other restrictions on access to the means of production or to the means of subsistence is to render the criteria for the identification of a mode of production, especially the concept of effective possession of the means of production, meaningless.

When the ideology and political practice of a "communal" society limits the access of certain categories of adults to the means of production, we must investigate the possibility that class relations are emerging or already exist in a previously unanalyzed form. Ideologically defended cultural practices such as the exchange of women by men (patriarchs arranging marriages for their daughters) can be considered compatible with communal relations of production only if women cannot be denied free access to the means of production should they refuse marriage. The exchange of women by men is neither "natural" nor necessary to the survival of lineage-based, subsistence societies as is sometimes alleged (Hindess and Hirst 1975; Gregory 1981). It is, however, highly conducive to the social reproduction of patriarchal relations of production.

The Class Structure of the Patriarchal Mode

In the patriarchal mode of production effective possession of the means of household-based production is monopolized by a class of patriarchs who are socially recognized as heads of household and/or extended family production units. The dependent class—wives, unmarried daughters, sons, and junior siblings of the patriarchal class—is denied free access to the means of production on the basis of ideological and political criteria which allow the patriarchal class to set the terms on which women and dependent males gain access to the means of production.

The patriarchal mode is characterized by an ideologically defined division of labor by sex, age, and family position which entails the performance of surplus labor by the subordinate class. The patriarchal class struggles to maintain control over the social definition of labor responsibilities and to retain exclusive authority to decide upon the ultimate distribution of the products of dependents' labor. Patriarchs may redistribute some of the products of dependents' labor, but to the extent that the dependent class works more hours than are embodied in the goods and services it consumes, it is engaging in surplus labor. (This abstracts from the problem of supporting children, the sick and the elderly, a problem which need not complicate the analysis if we assume that in an egalitarian situation, all adults would contribute equally to the consumption of family or community members who are incapable of full self support).

The class position of a patriarch is defined by control over economic resources essential to the daily and generational reproduction of all members of the production unit. The patriarch controls both the means of production and the means of family subsistence such as home-produced goods and services.[8] In a situation of articulating modes, depending on the relative strength of each mode and the state of the class struggle in the patriarchal mode, the patriarchal class may also control the incomes of male and/or female dependents from wage labor or petty commodity production.

Patriarchs control the generational reproduction of the production unit by restricting access to the sexual and childbearing capacities of female members. Women in the patriarchal mode do not control their own reproductive capacities. In societies where the mode is dominant, only patriarchs can exchange women in marriage arrangements. A dependent male cannot marry and thereby gain the social sanction to join the patriarchal class until some member of that class provides him with a wife.

Patriarchs establish and reproduce control over the means of production by engaging in class activities which foster the ideological legitimation and political consolidation of their effective control over economic resources. In precolonial Africa, for example, some patriarchal classes controlled the social rituals which were ideologically defined as preconditions for planting or harvesting. In most societies military and marriage decisions were considered the exclusive prerogatives of the patriarchal class (Laburthe-Tolra 1977; Marie 1976; Meillassoux 1964). Similar mechanisms of social control still exist in western societies in the form of male dominated religious and military hierarchies (Rubin 1975).

The unit of production in the patriarchal mode may be a single household or a group of households. The production unit may even vary by the nature of the activity (hunting in multihousehold groups, different types of farming in single and multihousehold groups), leading to differing specialized positions within the patriarchal class. Different strata of patriarchs may have different types or amounts of power (with heads of lineages, for example, dominating heads of households), but this variation does not negate the validity of the model just as the differentiation in economic and political power between oligopolists and small, competitive capitalists does not negate the validity of the capitalist mode as a theoretical construct. (When there is, however, a permanent form of social differentiation among patriarchs as when a class of feudal nobles or hereditary chiefs controls some of the means of production or distribution and extracts surplus labor both from the patriarchal class itself and from patriarchal dependents, the patriarchal mode is articulating with another mode in which both patriarchs and dependents have a second, subordinate class role.)

The subordinate class of the patriarchal mode has two distinct segments: females whose dependence on patriarchs is ideologically defined as permanent and males whose dependence is temporally limited to the early years of their life cycle. The female segment of the subordinate class is more thoroughly dominated and exploited than the male segment. To obtain access to the means of production (land, tools, cattle, male labor) and subsistence (monetary incomes, food, shelter), women are required to accept the labor obligations which patriarchal ideology and power attaches to the roles of daughter, wife, and mother. When those female labor obligations exceed the labor obligations of patriarchs, we have evidence of exploitation which benefits the patriarchal class.

Women can be both directly and indirectly exploited. They are directly exploited when their labor provides products and services which are immediately expropriated by the patriarchal class. They are indirectly exploited when they perform the childrearing work required for the survival of young children even though men control the labor and surplus labor of children who have reached a more productive age.[9]

Dependent males comprise the second segment of the subordinate class in the patriarchal mode. Male patriarchal dependents do not head their own households but are social and economic dependents of the patriarchal class. Dependent males obtain access to the means of production by performing production tasks and/or military service as directed by the patriarchal class.

Male dependents who are or become the socially legitimate sons or junior siblings of the patriarchal class are eventually liberated from dependent class status and integrated into the patriarchal class. The process of class transformation for dependent males is ideologically structured and socially confirmed in a series of rituals (initiation ceremonies, graduation exercises, marriage rites) which both effect and confirm the change in class status.

Because the process of male social emancipation affects crucial economic and political interests of the patriarchal class, it is tightly controlled by councils of elders, legislative bodies, and/or religious and military hierarchies which represent the interests of the patriarchal class as a whole.

Marriage is a central institution in the process of male class transformation. Before marriage a man has no means of creating his own sphere of dependents; his labor and its product are controlled by the patriarch upon whom he depends. Control over marriage arrangements gives patriarchs the means to control the labor of dependent males. On the other hand, patriarchal control of male labor power is temporally limited by the social expectation that patriarchs will provide their sons and junior brothers with wives before dependent males reach a certain age. Marriage then signals the beginning of a transformation period in which a man gains the right to dispose of the labor and reproductive potential of others, initially that of his first wife. As a man develops his own entourage of dependent wives and children, he becomes a full fledged member of the patriarchal class.

Marriage has no similar liberating significance for women. Instead a woman's marriage triggers a social expectation that she will perform the labor tasks ideologically defined as the obligations of a wife and mother. Furthermore, control over marriage requires patriarchal control over a very wide range of female behavior. Patriarchs can only procure wives for their male dependents if they have the power to supply wives to other members of the patriarchal class. Thus, control over marriage requires that patriarchs control their female dependents' place of residence, sexual activities, child-rearing activities, community of legitimate association, and labor power. The extensive nature of this type of social control requires a greater measure of oppression of female than male dependents. It also facilitates a greater degree of exploitation.

Male and female dependence, subordination, and exploitation in the patriarchal mode are structurally distinct. Male dependence is not only less oppressive and exploitative than female, it is temporary (except for male slaves). Female dependence, whether slave or not, is permanent. On a lifetime basis, males appropriate more surplus labor than they provide as dependents of the patriarchal class. Dependent (but free) males thus have little incentive to support women's class struggles because gains by women threaten their future class interests.

The dynamic of the patriarchal mode derives from the economic, political, and ideological advantages which accrue to patriarchs who increase the number of dependent persons in the production units they head. Dependents' labor produces the goods and services necessary to the biological reproduction of all members of a household—the patriarch, minor children and elderly non-working members and, of course, the laboring dependents themselves. In the patriarchal mode the accumulation of the material means of production as well as the means of human reproduction (subsistence products and women's childbearing capacities) takes place via the accumulation of human dependents. This aspect of the patriarchal mode helps explain the high rates of population growth in areas of the world where this mode still structures the social relations of production.

Articulation of Patriarchal and Non-patriarchal Modes

When the patriarchal mode is dominated by the feudal mode or by the capitalist mode, the ruling class of the dominant mode can expropriate surplus labor from the dominated mode either by bringing patriarchal dependents directly into the dominant mode (as wage or feudal laborers, for example) or by transferring the surplus labor of patriarchal dependents to the dominant mode in an indirect manner, as when dependents' labor is mobilized to produce a cash crop which patriarchs sell to merchant capitalists at a price which does not fully compensate for the labor embodied in the product.

In the current period, the patriarchal mode has been dominated in almost all societies by the capitalist mode. Even when dominated, however, the patriarchal mode can influence not only household production relations, but capitalist relations as well. When the patriarchal class retains access to land and other means of household production, allowing peasant families to produce most of their own subsistence needs and to withdraw from market-oriented production if commodity prices are very low, the reproduction of a patriarchal mode can impede the expansion of capitalist relations of production.

When capitalism extends its dominance to the point at which it is no longer possible for patriarchs to control sufficient economic resources to produce family subsistence requirements, patriarchs are forced to allow some members of patriarchal production units to sell their labor in the capitalist mode. But even though the patriarchal mode must then depend on its articulation with the capitalist mode to assure the survival of its dependent classes, patriarchal relations of production and distribution still result in patriarchal expropriation of surplus labor and continue to affect social ideology and political practice. Even when the patriarchal mode is highly dominated by capitalism as in western industrial societies, patriarchal attitudes and practice can affect capitalist development, as for example, when the division of labor in capitalist enterprises segments the majority of women into the lowest paid sectors of the work force. Patriarchal relations are also evident when women wage laborers face a "double day" of work at both the capitalist and household sites of production but male workers do not. Articulation does not, however, always depress the class position of a subordinate class. Dependent males or women in the patriarchal mode who simultaneously become wage workers often find their bargaining power in the patriarchal mode improved because of their access to non-patriarchal means of production. The discussion of articulation in Africa will illustrate this point.

The Patriarchal Mode in Africa

This section discusses the empirical features of a range of African precolonial societies to which an analysis guided by the model of a patriarchal mode of production can bring new insights and meaning. The discussion is not meant to suggest that the patriarchal mode is applicable to all African cases. The patriarchal mode should be regarded as one among two or three household based modes, such as the communal or the lineage modes, whose relevance and usefulness can only be determined in the course of an actual case study. I will argue, however, that the concept of a communal mode is often applied or assumed in cases where a more probing analysis would demonstrate that its basic assumptions are clearly violated. Initially section I presents examples of patriarchal class structure, social dynamics and class struggles from the precolonial period. I then discuss processes of articulation between patriarchal and capitalist classes during the colonial and postcolonial periods.

The Class Structure

Many marxian analyses of African precolonial societies describe the possession of land as communal. The role of male heads of households or lineages in managing economic resources, reproducing a particular sexual division of labor, controlling the production and distribution of food, and arranging marriages is presented as the community's way of assuring its survival rather than as a mechanism for expropriating surplus labor (Coquery-Vidrovitch 1976; Hymer 1970; Meillassoux 1978). This characterization of the patriarchs' role is unsatisfactory because it cannot explain the regular use of force which served to reproduce patriarchal domination of women and to secure their continual performance of surplus labor (Crehan 1984; Issacman 1978; Thiam 1978; Vincent 1976; Wagner 1970, cited in Hay 1982). Class contradictions between patriarchs and women are evident in the various forms of punishment patriarchs were socially expected to inflict upon women who failed to perform their socially defined obligations as wives. In many societies women who did not serve satisfying meals were subject to beatings. In some cases women who defied patriarchal constraints on their sexual activities risked torture or even death (Thiam 1978; Laburthe-Tolra 1976; Vincent 1976).

I reject the claim that land was possessed communally in cases where patriarchs could deny family members who defied patriarchal authority access to land. Patriarchal dependents were normally accorded access to land and other means of production in order to fulfill the economic roles expected of them, but this access was conditional and could be withdrawn. Women's access to the means of production was (and in most societies remains) predicated on their acceptance of a sexual division of labor in which the bulk of the work required to reproduce the members of the household was socially construed as the obligation of women as daughters, wives, and mothers. The specific sexual division of tasks between men and women varied among societies, but women were invariably expected to work considerably more hours than men when all the tasks necessary for the biological reproduction of the family are included.

In societies characterized by women's farming systems, "women's work" consisted of planting, weeding, harvesting, processing, storing, and cooking the food necessary to the daily sustenance of all members of the patriarchal unit (Boserup 1970; Bukh 1976). A woman was also expected to produce food surpluses beyond the needs of the household or lineage group in order to provide for her husband's guests and for the feasts which accompanied many ideological activities. In most societies women also produced household goods (baskets, pottery, etc.) and provided housekeeping and personal services for their husbands, children, and their husbands' guests and clients. Finally, all women were expected to bear and rear as many children as possible, a function which was critical to the enhancement of the power and prosperity of the patriarchal class.

Women's access to land depended on their acceptance of the sexual roles and labor obligations socially attributed to daughters and wives (Obbo 1980; Savané 1980; Strobel 1982). An unmarried girl's continued access to land depended on her willingness to accept a husband of her father's choosing and subsequently upon her ability to fulfill her obligations as wife and mother. If a woman found her marital situation intolerable and fled to her father's household, she was likely to be pressured or forced to return to her husband since the termination of a marriage could weaken an alliance between lineages or clans (Laburthe-Tolra 1977). If a wife persisted in her efforts to leave an estranged husband, she would be stripped of her land rights in the husband's village and forced to separate from her children (Cory and Hartnoll 1945; Goody and Buckley 1973; Hay 1982; Henn 1978). In order to activate land rights in her natal village, a divorced woman had to be accepted as a member of her father's or brother's household. If this was refused, her only means of survival was to find another patriarch willing to take her on as a wife or a slave (Albert 1971).

In most patrilineal societies, a young widow could only retain access to her food fields and to her children by marrying the man from her deceased husband's lineage who, in effect, inherited her (Guyer 1985). To refuse her inheritor was to risk banishment from her husband's village and loss of access to her children who "belonged" not to their mother, but to their father's lineage (Fortmann 1982; Obbo 1980).

A dependent man did not have access to land in his own right until his patriarch gave him a wife and the socially recognized right to begin accumulating his own set of dependents (Jewsiewsicki 1981; Laburthe-Tolra 1977; Marie 1976). During the period of their dependence, young men were obliged to perform agricultural or herding labor and military service in order to obtain food from their patriarch's wives. Prior to their emancipation, dependent males owed complete obedience to their patriarchs, but young men who were sons or nephews of patriarchs (rather than clients or slaves) could look forward to a period in which they too would have the right to control land, tools, and dependent labor. This meant that the class interests of dependent males were highly ambiguous; present and future class interests were even contradictory. Subordination in childhood and youth gave sons of the patriarchal class a class status similar to that of their mothers and sisters, but their socially required emancipation would eventually propel them into a dominant class position over all women including their mothers. To promote a resolution of the emotional ambivalence this type of life situation could produce, some male initiation cermonies forced young men to ideologically repudiate their mothers in order to affirm the change in their social relationship with women. One society even required male initiates to beat their mothers as testimony to their ability to disassociate themselves from the world of women (Brain 1978).

In most societies dependent males worked in agriculture and/or herding of livestock. In many parts of precolonial Africa, this work was supplemented or replaced by military service for the patriarch or, in societies where the patriarchal mode articulated with a feudal or other form of tributary mode, for the chief or king. In many societies effective possession of land required the allegiance and military strength of the class of dependent males. This was especially important in societies whose shifting agricultural cultivation required periodic migration to new areas and in pastoral societies which required access to huge tracts of land and several sources of water. Military prowess and the political and social prestige associated with it also facilitated the process of accumulating dependents. While the basic mechanism for enlarging one's sphere of dependents was the peaceful exchange of daughters as wives, in periods of demographic stress or when a lineage or clan was pursuing rapid expansion, raids and wars provided a means of obtaining women for whom no local dependents needed to be exchanged (Robertson and Klein 1983; Lovejoy 1981).

Mode of production analysis predicts that a dominant class's control over the means of production will result in the performance of surplus labor by a dependent class. If we accept the argument that average labor times in the patriarchal mode are socially necessary, historical records often provide the evidence necessary to make a quantitative estimate of surplus labor performance and/or expropriation. An estimate of women's surplus labor time and their rate of exploitation (the ratio of surplus to necessary labor time) can be made by summing the total labor hours performed in a typical production unit over a period of time, calculating the average number of hours each adult member of the unit would have to work if all labor were equally shared (as a proxy for necessary labor), and subtracting the proxy for necessary labor from women's actual labor time. For convenience of calculation, in the example which follows, the proxy for necessary labor is based on the hypothesis of equal consumption by all adults. For women this estimate of necessary labor may be biased upwards (and the rate of exploitation biased downward) because women often did not actually consume a product which embodied the average number of labor hours required of each adult for the society to have been equitably reproduced. If data is available, one should attempt to calculate necessary labor time correctly, e.g., on the basis of hours embodied in the goods and services women, dependent males, and patriarchs actually consumed.

Laburthe-Tolra's reconstruction of the pattern of daily activities of patriarchs, dependent males, and women in the Beti society of southern Cameroon during the late 19th century makes it possible to estimate surplus labor performance and consumption (1977, p. 652). Historical evidence indicates that women worked an average of 46 hours a week, dependent men 20 hours a week, and patriarchs 5 hours a week. Assuming that the average household production unit contained eight women (wives and daughters), six dependent males (sons and junior brothers or clients) and one patriarch, the reproduction of the household on a daily and generational basis if each adult had worked the same number of hours would have required adults to work 33 hours per week. Assuming equal consumption for men and women, 33 hours is our proxy for necessary labor. Women's surplus labor (actual less necessary labor) is thus 13 hours and their rate of exploitation (surplus/necessary labor) is 40 percent (13/33). Dependent males in this case were not exploited. Instead, they consumed 13 hours of women's surplus labor. Thus, the quantification of surplus and necessary labor highlights the ambiguity of the class position of dependent males in Beti society and also makes it abundantly clear that women were the major providers of surplus labor.[10]

The Social Dynamic

Many anthropological and historical studies of African stateless societies suggest that production for exchange was minimal in comparison with production for direct consumption (Terray 1972; Marie 1976). Important material surpluses seem mainly to have been produced in societies where the patriarchal mode was articulated with a tributary mode in which hereditary chiefs or kings used political and military institutions to extract surplus labor from their own subjects or from neighboring groups. In stateless societies, on the other hand, apart from the occasional production of luxury items like kola nuts and ivory for long distance trade, there was very little material surplus.

Some marxists have suggested that the absence of significant material surpluses is presumptive evidence of an absence of class contradictions and exploitation (Rey 1971; Terray 1972; Hindess and Hirst 1975). This conclusion fails to recognize that surplus labor need not take a product form. A great deal of women's surplus labor, for example, is in the form of childcare and domestic services. Failure to recognize the service forms of surplus labor reflects a failure to analyze the social relations of production within the main productive units of stateless societies—the households.

Low levels of material surplus do, however, suggest that the accumulation of goods was not the motivating force which explains patriarchal class behavior. What *did* motivate the patriarchal class? I propose that it was the accumulation of social dependents, specifically the wives, children, and other dependents who made up the subordinate class of the mode (cf. Jewsiewicki 1981). By accumulating dependents, patriarchs maximized their political power, social prestige, and material wealth. The labor power of dependents provided the patriarchal class with the control over products and services necessary to expand their households and lineages. Among the Gouro, for example, women not only produced the bulk of the household's food, childcare and other domestic services, they also produced the kola nuts which formed the basis of bridewealth (Meillassoux 1964; Marie 1976). Clearly dependents were both an immediate source of wealth and the means of further accumulation. But dependents in the patriarchal mode were even more the very substance of wealth itself. In many societies dependent women and children could be sent to a creditor's household as collateral for a loan or transferred to another patriarch permanently as payment for a fine or as a tribute (Bukh 1979; Hay 1982; Laburthe-Tolra 1977). In many societies, women, clients and slaves were the basis of a patriarch's inheritance.

Class Struggles

The record of women's resistence to patriarchal domination in precolonial Africa is based primarily on stories and oral histories indicating that tactics of resistance and defiance were largely individual, although women did count on assistance from other women to help hide illicit affairs, to defuse the wrath of a demanding husband, or even to flee from a husband's village in order to seek refuge with a father, brother, or lover (de The 1970; Vincent 1976; Thiam 1978).

For women, united class action was especially difficult in patrilineal, patrilocal societies where nearly all adult males were linked by kinship, but most women and all wives were strangers from unrelated lineages. Female solidarity had to be built among women of varying cultural backgrounds and social positions. Upon entering her husband's household (often as early as seven years of age), a young girl often faced an internal hierarchy of senior, junior, and favorite wives. These divisions in the female component of the dependent class prompted women to concentrate on their personal situations rather than on female class action. A woman who faithfully fulfilled her obligations as wife and mother could improve her social status and eventually reduce her surplus labor obligations. Senior wives, for example, might be exempted from heavy agricultural work if their husbands had junior wives. There was also the possibility of becoming a favorite wife by pleasing one's husband economically, socially, or sexually. In some cases, however, social custom served to thwart women's personal efforts to attenuate their exploitation. Among the Beti, for example, when a patriarch died his favorite wife was usually among the widows who were immediately executed by his sisters' sons as social retribution for his death (Laburthe-Tolra 1977, pp. 1283–1296). I would interpret this custom as a warning from the patriarchal class to all women that it was extremely dangerous to try to overcome their subordinate status.

The Beti case may, however, represent an exceptionally repressive example of the patriarchal mode. In contrast, political and ideological structures in some precolonial societies provided relatively advantaged positions for certain women. For example, the institution of woman-to-woman marriage in a few societies allowed a widow or even a wealthy married woman to pay another woman's brideprice and take a patriarchal class position vis-a-vis her "wife" (Sacks 1982, 77–79; O'Brien 1977). In some societies, widows and divorced women could become the heads of their own households and even arrange marriages and assume male political roles for the household (MacCormack 1982). More commonly, parallel male and female political and ideological roles were developed in which women led the female portion of the community (Okonjo 1976; Lebeuf 1963). Such roles for women can be variously interpreted as the means by which the patriarchal class coopted particularly capable and potentially rebellious women, as successful examples of women's resistance to male control, or as evidence which invalidates the relevance of the patriarchal mode. While only a comprehensive case study can establish the plausibility of any of these interpretations in specific situations, I would nonetheless suggest that a theoretically rigorous analysis of the social relations of household production would probably show that most societies with some favored positions for women were still predominantly patriarchal. The existence of limited political and ideological leadership roles for women does not *ipso facto* invalidate the general relevance of the patriarchal mode of production.

Hindess and Hirst (1975, p. 78) have argued that patriarchs had no class-like power over either dependent males or women because any dissatisfied member of a household or lineage could simply "vote with his feet" and then presumably attach him or herself to another household or lineage. There is little evidence to support this assertion. For a young man to leave his patriarch was to jeopardize his opportunity to become a member of the patriarchal class (Laburthe-Tolra 1977; Terray 1972; Marie 1976). Men who voluntarily left home or whose lineage was defeated militarily were often forced to become permanent dependents, i.e., clients, or in the worst case, slaves. When women fled to their natal villages, inter-lineage social arrangements among patriarchs would be activated to return them. Among the Beti, for example:

> The drums immediately spread the news of the flight of a woman; any man giving her refuge without returning her immediately to her husband-owner was *ipso-facto* guilty of adultery (Laburthe-Tolra 1977, p. 557).

Women who dared to wander about the forest without male protection were subject to kidnapping and enslavement by more powerful lineages (Wright 1975; Robertson and Klein 1983). It seems clear that patriarchs understood the need to take united class action to thwart the rebellious class struggles of their dependents.

It was primarily during the colonial period, when the patriarchal mode was forced to accommodate itself to articulation with the foreign capitalist mode, that flight became a more viable way for dependents to opt out of a particular patriarchal household and seek a better situation. Dependent males could flee to colonial plantations or towns to seek work, while women could seek refuge in Christian missions (de The 1970; Beti 1971).

Patriarchy and Capitalism

The process of articulation of capitalist and noncapitalist modes in Africa has been marked by force, mutual accommodation, and a complex mix of gains and losses for people caught up in the changing class relationships. In the colonial period, foreigners assumed the role of capitalists and created the colonial state. In most cases, however, the African patriarchal class did not lose its access to land and could therefore retain its ability to reproduce the household production unit. Because the autonomy of the patriarchal mode made it difficult for colonial capitalists to obtain sufficient supplies of cheap labor for profitable operation, the foreign capitalist class and its colonial state devised other ways to penetrate the African household. Dependent males were coerced into forced labor projects, women were required to produce food for colonial enterprises, and patriarchs were obliged to produce the export crops demanded in European markets (Brown 1980; Guyer 1980; Hay 1976; Stichter 1975-76).

Dependent males were the first class category from the patriarchal mode to occupy working class positions in capitalist enterprises. Although they had initially been forced out of the patriarchal household, dependent males sometimes found that wage labor provided an alternative to labour for patriarchs who were attempting to extract more surplus labor from dependents in response to colonial pressure to increase export crop production. As the growing availability of capitalist and state employment partially eroded patriarchal ability to control dependent *male* labor, patriarchs demanded more surplus labor from women (Bukh 1979; Chanock 1982).

Women's intensified surplus labor was directly appropriated in the patriarchal mode and indirectly transferred to the capitalist mode. For example, when women's food products were given without payment to sons and brothers working in capitalist enterprises or at forced labor sites, capitalists who were directly exploiting African men were also indirectly exploiting women (Guyer 1980; Stichter 1975-76).

The increasing exploitation of women during the colonial period sometimes brought benefits to dependent males as well as to capitalists. By consuming the products of his mother's surplus labor, a wage earning son who was a dependent in the patriarchal mode might be able to accumulate a substantial part of the bridewealth necessary to secure his first wife, thereby accelerating his entry into the patriarchal class. The son's gain represented a direct loss to the mother. Mothers who continued to feed sons residing outside the village had to substitute their own increased labor on field clearing and other agricultural tasks for labor previously performed by dependent males. At the same time, women farmers also faced increasing colonial demands to produce food for forced labor camps, state administrative centers, and the caravans of human porters passing through their villages (Guyer 1980; Henn 1978).

On the other hand, the penetration of colonial capitalism also provided women with new opportunities to struggle against patriarchal control and oppression. The gradual development of colonial markets for food represented an opportunity to sell food for cash. Among the Yoruba and Ga peoples of West Africa, much of the regional food and other retail trade was carried on by women who thereby gained access to means of distribution and sources of supply which were not entirely controlled by the patriarchal class (Robertson 1984). Still, women's cash earnings were almost never sufficient to free them from patriarchal dependence, i.e. pay back their brideprice and to support themselves and their children.

Within the patriarchal household, most women faced a struggle with husbands over the disposition of their earnings from commodity production or trade. Patriarchs whose traditional prerogatives had included control over the proceeds of women's subsistence labor often asserted their "right" to receive women's cash earnings (Kayberry 1952; Mayer 1950; cited in Hay 1982; Obbo 1980).

In general, women's access to class roles outside the patriarchal mode was much more limited than that of dependent males. In part this was due to an implicit agreement between the foreign capitalist class and the colonial state, on the one hand, and the African patriarchal class, on the other, that patriarchs should retain direct control of women's labor as long as some of the product of women's surplus labor was transferred to the capitalist sector. Accordingly, colonial capitalists rarely hired women as wage laborers. But neither capitalist nor patriarchal class action could completely choke off women's migration to colonial towns where some women survived by producing cooked food, beer, and sexual services for male wage earners (Dugast 1955–56; Muntemba 1982; Stichter 1985; Strobel 1979).

Female migration aroused fierce opposition on the part of the patriarchal class. In parts of East Africa, a woman who migrated to town on her own volition had to endure social ostracism as a "prostitute" for the rest of her days (Storgaard 1975–76). Patriarchs continually appealed to colonial authorities to use the power of the state to help keep women under control. Recognizing the centrality of women's labor to the creation of an exploitable surplus on African farms, the colonial authorities responded with legislative and administrative measures to strengthen patriarchal dominance. From time to time, for example, single women living in colonial towns were rounded up by the police and forcibly returned to their villages (Chauncey 1981; Chanock 1982; Jacobs 1984; Strobel 1979).

For patriarchs, capitalist penetration of the household and the society produced cross-cutting effects. Patriarchs clearly lost political and military power as a result of colonial conquest. Their ideological supremacy and their ability to impose life-threatening sanctions on dependents was diminished by the colonial state. Patriarchal control over dependent males was fundamentally threatened by the existence of capitalist wage labor as an alternative means of subsistence. On the other hand, for some members of the patriarchal class, cooperation with colonial authorities brought unprecedented increases in material wealth. Economic and political power differentials among patriarchs were greatly exacerbated by colonial enrichment of those members of the patriarchal class who facilitated in the recruitment of forced labor and the collection of colonial taxes (Jewsiewicki 1981; Laburthe-Tolra 1977).

The effects of the articulation of capitalist and patriarchal modes are still evident. If, as argued above the dynamic force of the patriarchal mode prior to its articulation with capitalism was the patriarchal motivation to accumulate dependents, capitalist penetration has clearly not destroyed the powerful ideology which reinforced that dynamic: for many Africans "real wealth" is still wealth in people. People's desire to have a large number of children results in part, but not entirely, from the continuing ability of the patriarchal class to command economic services and monetary support from their children. The economic benefit of dependents is most apparent in the rural areas where the patriarchal class still controls both land and the simple tools required for traditional forms of agricultural production. In urban areas on the other hand, it is often extremely difficult to support a large number of dependents unless the patriarch is a capitalist, a state bureaucrat, or a member of the small labor aristocracy. Still, many poor urban parents often have large families, a phenomenon which may well reflect the contemporary impact of the traditional patriarchal belief that sexual virility must be continually demonstrated.

Finally the predominance of patriarchal social relations of production in African rural economies has fundamental implications for the nature and possibilities of capitalist development itself. As the capitalist dominated state continues to extract surplus from the rural sector through its control over agricultural commodity markets and its penetration of the process of cash crop production, patriarchs are exploited in a manner which directly limits their monetary incomes but has less of a negative effect on their material welfare as a class because patriarchs can use their own class power to pass much of the burden of producing the rural surplus on to women and dependent men. Male patriarchal dependents attempt to reduce the negative effect of patriarchal exploitation on their own material welfare by seeking wage employment in the capitalist mode, thus forming the basis of capital's huge reserve of cheap labor, the plethora of people for whom the very low level of wages in even the most modern of internationally managed firms is superior to their alternatives in the dominated patriarchal mode. Women in the patriarchal mode can be seen to bear the greatest burden of this articulation of state capitalist and patriarchal exploitation when account is taken of the surplus labor they perform in the subsistence sector.

Capitalist domination of the patriarchal mode depresses African aggregate demand for capitalist output, while the patriarchal division of labor which holds women responsible for subsistence food and services and allows men to concentrate on marketable commodities skews the demand for capitalist produced commodities toward items of primary interest to men. Women's effective demand in capitalist markets is far too marginal, for example, to stimulate much capitalist investment in industries producing tools or inputs specific to food farming or in labor-saving food processing equipment. On the other hand, it is no accident that the most profitable consumer goods industries in most African countries are beer and cigarettes.

To summarize, the model of articulating capitalist and patriarchal modes of production helps explain not only the continuing ability of African capitalist enterprises to attract workers at extremely low wage levels, but also the very limited pace of African capitalist development and its continuing orientation toward exports and male preferred commodities.

Conclusion

This paper has attempted to demonstrate that an articulation of modes of production analysis which incorporates a concept of the patriarchal mode of production places the issues of women's subordination and the inter-relations between class and gender in capitalist dominated societies in a new analytical framework—a framework which poses new questions and throws new light on old debates. The debate over the origin and intensification of women's subordination in Africa is a case in point. Some participants in this debate have argued that capitalist penetration of African economies was the basic cause of women's subordination, while others contend that women's social and economic choices were even more limited in the precapitalist period than they are today. Empirical evidence relevant to this debate highlights the ambiguity of the analytical concepts of gender analysis. On the one hand, women's labor time has increased substantially since the intensification of capitalist penetration during the colonial period with little or no positive impact on most women's material welfare. (This is also true, though to a lesser extent, for men). On the other hand, women's ability to influence decisions as to whom they will marry or whether or not they will divorce (choices which for most women determine where and with whom they will work) is definitely greater today than it was in the precolonial period. What the totality of such historically relevant evidence means for "women's subordination" is not at all clear.

An historical analysis of the changes brought about by the articulation of precapitalist and capitalist modes can place the subordination debate in a new perspective. Use of the mode of production framework both forces and facilitates an analysis of the nature and interaction of the economic, political, and ideological characteristics of all historically relevant modes of production. When such an analysis results in significant evidence of women's economic exploitation and political and ideological subordination, the in-corporation of the patriarchal mode of production into an articulation of modes of production analysis can not only help improve our understanding of the determinants and effects of women's subordination in both precapitalist and capitalist periods, but it can also facilitate a politically relevant analysis of the intimate interrelations between capitalist and patriarchal class struggles.

Class analysis is clearly much more complex when we theoretically address the possibility that individuals participate in more than a single class relation.

But if this complexity is a fundamental aspect of reality, we cannot produce a relevant analysis by ignoring it. If our analysis leads to the conclusion that a patriarchal mode of production explains basic aspects of the social reality we are trying to understand, we simply must recognize that, in that case, gender itself has become the basis for class.

Notes

1. I first attempted to define a patriarchal mode of production in Henn, 1978. In 1983 Nancy Folbre and I discovered that we had both been working on essentially the same concept. Since that time we have shared our work. I have benefitted enormously from this exchange. See especially Folbre 1982, 1983, 1986.

2. See Foster-Carter 1978, and Wolpe 1980, for reviews and critiques of the modes of production literature; Katz 1980, for an extended treatment of the development and application of the theory in African studies; Crummey and Stewart 1981, and Seddon 1978, for African case studies; and the *Canadian Journal of African Studies* Vol. 19, No. 1 (1985) for the views of 19 Africanist scholars on the uses and usefulness of the theory in African studies.

Slavery and pawnship have not been incorporated into this argument. To do so would require another study, with different case material. Considering the spate of excellent work on slavery and pawnship in recent years (Lovejoy, 1981 and 1983; Robertson and Klein, 1983), such a study would be both possible and important.

3. Protagonists in the domestic labor debate tried to bring housework and childrearing labor into the sphere of Marxian value analysis by arguing that housewives' unpaid labor reduces the value of labor power (the "commodity" workers sell to capital for a wage) and thus cheapens the cost of wage labor to capital (Dalla Costa and James 1970; Harrison 1973; Zaretsky 1973; Seccombe 1974). The approach was appealing because it moved beyond Marx's implausible assumption that the value of labor power (defined as the labor time required to reproduce labor power) is equivalent to the labor time embodied in the commodities purchased with the wage. The problem with Marx's formulation, of course, is that all the labor of childrearing, cooking, etc., which clearly goes into the reproduction of labor power, is ignored, while commodities purchased with the wage are consumed by family members as well as the wage earner.

The critics in the domestic labor debate insisted that "concrete labor in the domestic sphere and the abstract labor of commodity production" are not commensurable, "hence there is no basis for the calculation of a transfer of surplus-labor time between the two spheres" (Molyneux 1979, p. 9; see also Himmelweit and Mohun 1977). This position seems to have largely ended the debate among the original participants. Some new contributions to Marxian theory may reopen the discussion. See Bryceson 1983, for a rigorous, theoretical refutation of incommensurability between abstract and concrete labor.

4. Folbre (1986) demonstrates the relevance of the patriarchal model to the analysis of both Western and Third World societies.

5. A brief overview of this literature which includes most of the recent contributions is found in the introduction to the *Review of Radical Political Economics'* special

issue, "Modern Approaches to the Theory of Value" 14: 2 (Summer 1982) pp. 1–5.

6. This is not to deny Marx's insight that each mode of production which he identified theoretically—the primitive communist, ancient, Asiatic, feudal, and capitalist—would, at the peak of its development, have a predictable correspondence between the forces and the social relations of production (Marx 1965, pp. 95–96). Nevertheless, when one is employing mode of production concepts for the analysis of a social formation which contains several modes of production in various stages of transition and articulation with one another, one should not expect to find a unique correspondence between the social relations and the forces of production for any one mode.

7. As Harold Wolpe points out, some of the influential writers on mode of production theory (Hindess and Hirst 1975; Laclau 1971) have restricted the concept of the economy in a mode of production to the combination of the relations and forces of production without specifying the "mechanisms of reproduction or laws of motion of the 'economy' as a whole" (Wolpe 1980, p. 7). I argue that an understanding of the mechnisms of reproduction of the social relations of production requires analysis of the political and ideological factors which make this social reproduction of economic relationships possible. In addition, the biological reproduction of the labor force and the physical reproduction of the means of production require an understanding of the ideological and political factors affecting the processes of distribution, consumption, and accumulation.

8. The means of subsistence denote all the consumption goods and services by which the members of a household are biologically reproduced on a daily and on a generational basis. Where the patriarchal mode articulates with capitalism, wage earnings and/or petty commodity earnings may enter the means of subsistence. Sexual services, childbearing and childrearing enter this category in all manifestations of the patriarchal mode. In recent socialist feminist literature the means of subsistence are sometimes referred to as the means of reproduction.

9. Many African customary laws which regulate parental custody over children in cases of divorce provide examples of the indirect form of exploitation. Children who are under the age of seven are normally allowed to remain with their mothers, but children over the age of seven have traditionally been awarded to their fathers. It is very interesting to note that labor time studies which have included children have found that seven is approximately the age when children begin to contribute significant amounts of domestic and agricultural labor to African rural households (Caldwell 1982).

10. This calculation is complicated by lack of data on the number of hours dependent males spent on military training and service. They would have to serve over 20 hours of military labor a week to arrive at a point where their actual labor equalled necessary labor.

Acknowledgments

This paper has been so long in the gestation stage that a proper acknowledgement of the many friends who gave time and thought to its improvement is impossible. I am especially grateful to colleagues from the New York City and Amherst, Massachusetts Women and Development study groups who critically discussed earlier versions. Colleagues at the Bunting Institute of Radcliffe College also helped sharpen the arguments even when they disagreed. Throughout the process Nancy Folbre provided exceptional insights, Jeanne Pyle offered extremely thoughtful comments and Sharon Stichter gave me the courage to "deliver."

References

Albert, E.M. (1971) "Women of Burundi: A Study of Social Values," pp. 179–215 in D. Paulme (ed.), *Women of Tropical Africa*. Berkeley: University of California Press.

Balibar, E. (1970) "The Basic Concepts of Historical Materialism," pp. 199–308 in L. Althusser and E. Balibar, *Reading Capital*. London: New Left Books.

Banaji, J. (1977) "Modes of Production in a Materialist Conception of History," *Capital and Class* 3: 1–44.

Beneria, L. (1979) "Reproduction, Production and the Sexual Division of Labour," *Cambridge Journal of Economics* 3: 203–225.

Bernstein, H. (1979) "African Peasantries: a Theoretical Framework," *Journal of Peasant Studies* 6(4): 421–443.

Beti, M. (1971) *The Poor Christ of Bomba*. London: Heinemann.

Boserup, E. (1970) *Woman's Role in Economic Development*. London: Allen & Unwin.

Bowles, S. and Gintis, H. (1982) "On the Class-Exploitation-Domination Reduction," *Politics and Society* 11(3).

Brain, J. (1978) "Symbolic Rebirth: The Mwali Rite among the Luguru of Eastern Tanzania," *Africa* 48(2): 176–88

Brown, B.B. (1980) "Women, Migrant Labor and Social Change in Botswana." Working paper no. 41, African Studies Center, Boston University.

Bryceson, D.F. (1983) "Use Values, the Law of Value and the Analysis of Non-capitalist Production," *Capital and Class* 20: 29–63.

Bryceson, D.F. and Vuorela, U. (1984) "Outside the Domestic Labor Debate: Towards a Theory of Modes of Human Reproduction," *Review of Radical Political Economics* 16 (2/3): 137–166.

Bukh, J. (1979) *The Village Women in Ghana*. Uppsala: Scandinavian Institute of African Studies.

Caldwell, J. (1982) *Theory of Fertility Decline*. New York: Academic Press.

Chanock, M. (1982) "Making Customary Law: Men, Women and Courts in Colonial Northern Rhodesia," pp. 53–67 in M.J. Hay and M. Wright (eds.), *African Women and the Law: Historical Perspectives*. Boston: Boston University Press.

Chauncey, G. Jr. (1981) "The Locus of Reproduction: Women's Labour in the Zambian Copperbelt, 1927–1953," *Journal of Southern African Studies* 7(2): 135–64.

Coquery-Vidrovitch, C. (1976) "The Political Economy of the African Peasantry and Modes of Production," pp. 90–111 in P.C.W. Gutkind and I. Wallerstein (eds.), *The Political Economy of Contemporary Africa*. London: Sage.

Cory, H. and Hartnoll, H.H. (1945) *Customary Law of the Haya Tribe, Tanganyika Territory.* London: Lund.

Crehan, K. (1984) "Women and Development in Northwestern Zambia: From Producer to Housewife," *Review of African Political Economy* 27/28: 51–66.

Crummey, D. and Stewart, C.C. (eds.). (1981) *Modes of Production in Africa: the Pre-Colonial Era.* Beverly Hills/London: Sage.

Dalla Costa, M. and James, S. (1970) *The Power of Women and Subversion of the Community.* Bristol: Falling Wall Press.

de The, M.P. (1970) "Influence des femmes sur l'evolution des structures sociales chez les Beti." These pour le doctorat de 3eme cycle. Ecole Pratique des Hautes-Etudes. Paris.

Dugast, I. 1955–56 *Monographie de la tribu des Ndiki: Banen du Cameroun.* 2 vols. Paris: Institut d'Ethnologie.

Folbre, N. (1982) "Exploitation Comes Home: A Critique of the Marxian Theory of Family Labour," *Cambridge Journal of Economics* 6:317–329.

———. (1983) "Of Patriarchy Born: The Political Economy of Fertility Decisions," *Feminist Studies* 9(2); 261–284.

———. (1986) "A Patriarchal Mode of Production," In R. Albelda, (ed.) *New Directions in Political Economy.* New York: M.E. Sharpe.

Fortmann, L. (1982) "Women's Work in a Communal Setting: the Tanzanian Policy of Ujamaa," pp. 191–205 in E.G. Bay, (ed.), *Women and Work in Africa.* Boulder, Colorado: Westview Press.

Foster-Carter, A. (1978) "The Modes of Production Controversy, " *New Left Review* 107:47–77.

Goody, J. and Buckley, J. (1973) "Inheritance and Women's Labor in Africa," *Africa* 43(2): 108–21.

Gregory, C.A. (1981) "A Conceptual Analysis of a Non-capitalist Gift Economy with Particular Reference to Papua New Guinea," *Cambridge Journal of Economics* 5:119–135.

Guyer, J.I. (1980) "Female Farming and the Evolution of Food Production Patterns amongst the Beti of South Central Cameroon," *Africa* 50(4): 341–56.

———. (1985) "The Economic Position of Beti Widows, Past and Present," pp. 313–325 in J.C. Barbier, (ed.), *Femmes du Cameroun: Meres Pacifiques, Femmes Rebelles.* Paris: Orstom/Karthala.

Harrison, J. (1973) "The Political Economy of Housework," *Bulletin of the Conference of Socialist Economists* 4(1).

Hartmann, H. (1979) "Capitalism, Patriarchy, and Job Segregation by Sex," pp. 206–47 in Z.R. Eisenstein, (ed.), *Capitalist Patriarchy and the Case for Socialist Feminism.* New York: Monthly Review Press.

Hay, M.J. (1976) "Luo Women and Economic Change during the Colonial Period," pp. 87–109 in N.J. Hafkin and E.G. Bay, (eds.), *Women in Africa.* Stanford: Stanford University Press.

———. (1982) "Women as Owners, Occupants, and Managers of Property in Colonial Western Kenya," pp. 110–123 in M.J. Hay and M. Wright (eds.), *African Women and the Law: Historical Perspectives.* Boston: Boston University Press.

Henn, J.K. (1978) "Peasants, Workers and Capital: The Political Economy of Labor and Incomes in Cameroon." Ph.D. Dissertation, Harvard University.

———. (1985) "Economic Ties between Peasant and Worker," pp. 393–400 in J.C. Barbier, (ed.), *Femmes du Cameroun: Meres Pacifiques, Femmes Rebelles*. Paris: Orstom/Karthala.

Himmelweit, S. and Mohun, S. (1977) "Domestic Labour and Capital," *Cambridge Journal of Economics* 1:15–31.

Hindess, B. and Hirst, P. (1975) *Pre-Capitalist Modes of Production*. Boston: Routledge and Kegan Paul.

———. (1977a) *Mode of Production and Social Formation*. London: Macmillan.

———. (1977b) "Mode of Production and Social Formation in PCMP: A Reply to John Taylor," *Critique of Anthropology* 8:49–58.

Hymer, S. (1970) "Economic Forms in Precolonial Ghana," *Journal of Economic History* 20 (1):33–50.

Issacman, A. (1978) *A Luta Continua: Creating a New Society in Mozambique*. Binghamton, New York: Braudel Center.

Jacobs, S. (1984) "Women and Land Resettlement in Zimbabwe," *The Review of African Political Economy* 27/28:33–50.

Jewsiewicki, B. (1981) "Lineage Mode of Production: Social Inequalities in Equatorial Central Africa," pp. 93–114 in Donald Crummey and C.C. Stewart (eds.), *Modes of Production in Africa*. London: Sage.

Katz, S. (1980) *Marxism, Africa, and Social Class*. Montreal, Quebec: McGill University.

Kayberry, P.M. (1952) *Women of the Grassfields*. London: Her Majesty's Stationary Office.

Laburthe-Tolra, P. (1977) *Minlaaba: Histoire et societe traditionelle chez les Beti du Sud Cameroun*. Paris: Librairie Honore Champion.

Laclau, E. (1971) "Feudalism and Capitalism in Latin America," *New Left Review* 67:19–38.

———. (1977) *Politics and Ideology in Marxist Theory*. London: New Left Books.

Lebeuf, A.M.D. (1963) "The Role of Women in the Political Organization of African Societies," pp. 93–120 in D. Paulme, (ed.), *Women of Tropical Africa*. Boston: Routledge and Kegan Paul.

Lovejoy, P. ed. (1981) *The Ideology of Slavery in Africa*. Beverly Hills: Sage.

Lovejoy, P. (1983) *Transformation in Slavery. A History of Slavery in Africa*. New York: Cambridge University Press.

MacCormack, C.P. (1982) "Control of Land, Labor, and Capital in Rural Southern Sierra Leone," pp. 35–33 in E.G. Bay, (ed.), *Women and Work in Africa*. Boulder, Colorado: Westview Press.

MacFarlane, A. (1978) "Modes of Reproduction," *Journal of Development Studies* 14(4): 100–120.

Marie, A. (1976) "Rapports de parente et rapports de production dans les societes lignageres," pp. 86–116 in Francois Pouillon (ed.). *L'Anthropologie Economique: Courants et Problemes*. Paris: Maspero

Marx, K. (1906) *Capital: A Critique of Political Economy*. New York: The Modern Library.

———. (1967) *Capital*. Vol. III. New York: International Publishers.

———. (1965) *Pre-capitalist Economic Formations*. New York: International Publishers.

Mayer, P. (1950) *Gusii Bridewealth, Law and Custom*. Oxford: Oxford University Press.

Meillassoux, C. (1964) *Anthropologie economique des Gouro de Cote d'Ivoire*. Paris: Mouton.

———. (1975) *Femmes, Greniers, et Capitaux*. Paris: Maspero.

———. (1978) "The 'Economy' in Agricultural Self-Sustaining Societies: A Preliminary Analysis," pp. 127–57 in David Seddon (ed.), *Relations of Production: Marxist Approaches to Economic Anthropology*. London: Frank Cass.

Molyneux, M. (1979) "Beyond the Domestic Labour Debate," *New Left Review* 116:3–27.

Muntemba, M.S. (1982) "Women and Agricultural Change in the Railway Region of Zambia: Dispossession and Counterstrategies, 1930–1970," pp. 83–103 in E.G. Bay, (ed.), *Women and Work in Africa*. Boulder: Westview Press.

Obbo, C. (1980) *African Women: Their Struggle for Economic Independence*. London: Zed Press.

O'Brien, D. (1977) "Female Husbands in Southern Bantu Societies," pp. 109–126 in A. Schlegel, (ed.), *Sexual Stratification: A Cross-cultural View*. New York: Columbia University Press.

Okonjo, K. (1976) "The Dual-Sex Political System in Operation: Igbo Women and Community Politics in Midwestern Nigeria," pp. 45–58 in N.J. Hafkin and E.G. Bay, (eds.), *Women in Africa*. Stanford: Stanford University Press.

Resnick S. and Wolff, R.D. (1982) "Classes in Marxian Theory," *Review of Radical Political Economics* 13(4): 1–18.

Rey, P.P. (1971) *Colonialisme, Neo-Colonialisme et Transition au Capitalisme: Example de la "Comilog" au Congo-Brazzaville*. Paris: Maspero.

———. (1973) *Les alliance de Classe*. Paris: Maspero.

———. (1979) "Class Contradiction in Lineage Societies," *Critique of Anthropology* 13/14: 41–60.

Roberts, P. (1984) "Feminism in Africa: Feminism and Africa," *The Review of African Political Economy* 27/28: 175–84.

Robertson, C.C. (1984) *Sharing the Same Bowl*. Bloomington: Indiana University Press.

Robertson, C.C. and Klein, M.A. (eds.) (1983) *Women and Slavery in Africa*. Madison: University of Wisconsin Press.

Rubin, G. (1975) "The Traffic in Women: Notes on the 'Political Economy' of Sex," pp. 157–210 in R. Reiter, (ed.), *Toward an Anthropology of Women*. New York: Monthly Review.

Sacks, K. (1982) *Sisters and Wives: The Past and Future of Sexual Equality*. Chicago: University of Illinois Press.

Savane, M.A. (1980) "Women and Rural Development in Africa," pp. 26–32 in *Women in Rural Development: Critical Issues*. Geneva: International Labour Office.

Seccombe, W. (1974) "The Housewife and Her Labor Under Capitalism," *New Left Review* 83:3–24.

Seddon, D., (ed.) (1978) *Relations of Production: Marxist Approaches to Economic Anthropology*. London: Frank Cass.

Stichter, S. (1975–76) "Women and the Labor Force in Kenya, 1895–1964," *Rural Africana* 29:45–67.

———. (1985) *Migrant Labourers*. Cambridge: Cambridge University Press.

Storgaard, B. (1975-76) "Women in Ujamaa Villages," *Rural Africana* 29:135-153.

Strobel, M. (1979) *Muslim Women in Mombasa 1890-1975.* New Haven: Yale University Press.

———. (1982) "African Women," *Signs* 8(1): 109-131.

Taylor, J. (1975) "Pre-capitalist Modes of Production," *Critique of Anthropology* 4/5: 127-142 and 6: 56-69.

Terray, E. (1972) *Marxism and "Primitive" Societies.* New York: Monthly Review Press.

Thiam, A. (1978) *La parole aux negresses.* Paris: Editions Denoel/Gonthier.

Vincent, J.F. (1976) *Traditions et Transitions: Entretiens avec des Femmes Beti du Sud-Cameroun.* Paris: Berger-Levrault.

Wolpe, H. (1980) "Introduction," pp. 1-43 in H. Wolpe, (ed.), *The Articulation of Modes of Production.* London: Routledge and Kegan Paul.

Wright, M. (1975) "Women in Peril: A Commentary upon the Life Stories of Captives in Nineteenth Century East-Central Africa," *African Social Research* 20: 800-891.

Zaretsky, E. (1973) *Capitalism, the Family and Personal Life.* New York: Harper-Colophon.

3

Patriarchal Social Formations in Zimbabwe

Nancy Folbre

Until recently, economic research on women in Africa has focused largely on their hitherto neglected role in economic development, filling in blank spaces on an intellectual map whose peripheral regions remained unexplored (Boserup 1970). As the blank spaces have gradually filled, however, the traditional boundaries have begun to seem suspect. Though they once appeared peripheral, gender and age-based inequalities are emerging as central features in the larger economic and political landscape traditionally mapped in terms of inequalities based on nation, race, and class.

Now, the conceptual boundaries are being re-drawn. Elsewhere in this volume, Jeanne Henn locates the subordination of women in the central concept of a patriarchal mode of production. Like Henn, I argue that patriarchy is neither an aspect of capitalism nor an autonomous sytem, but rather a mode of production in patriarchal social formations. My conceptualization grows out of a feminist critique and revision of several Marxian concepts: a widening of the term production to encompass childbearing and childrearing, and a focus on the social relations that govern population growth as well as those that govern capital accumulation (Folbre 1986). My intent is not to reduce gender inequality to its economic dimensions but rather to reveal the unique logic of a multidimensional mode of production,

This paper benefited from related research funded by the Beijer Institute of the Royal Swedish Academy of Science in Zimbabwe in 1983. I gratefully acknowledge the helpful comments and criticisms of Tom Harris, Dan Weiner, Kirsten Johnson, Bonnie Ram, Pauline Ong, and Barry Munslow, while absolving them of any responsibility for the final results. Jeanne Henn and other members of the New England Women in Development group helped me develop my ideas regarding patriarchal social formations. Special thanks to Sharon Stichter, whose encouragement and careful editing were indispensable.

to emphasize the complex and contradictory interaction between distinct modes of production rather than the *a priori* dominance of one over others.

In this paper, I develop a new interpretation of the political and economic evolution of Zimbabwean society that places patriarchal social relations at the very center of a larger explanation of the establishment, consolidation, and breakdown of colonial control. Four related points pertain to different periods of Zimbabwean history: (1) The society that British colonizers encountered in Zimbabwe can best be characterized as a patriarchal tributary social formation. Patriarchal social relations among both the Shona and Ndebele that enabled men to extract a surplus from women and children were overlaid by tributary relations that funneled surplus to an Ndebele military elite. (2) After 1886, the Rhodesian Native Reserves system not only served the interests of white employers but also helped African men maintain patriarchal control over women. The persistence of the reserves system represented a concession to the patriarchal mode of production that may have inhibited the full development of capitalist relations. (3) The rapid population growth that contributed to the destabilization of colonial rule after 1940 was an important outcome of the juxtaposition of strong patriarchal social relations in the native reserves with new health technologies introduced by colonial capital. The persistence of traditional patriarchal relations also helps explain why young people played a particularly important role in the liberation struggle that intensified in the 1970s. (4) Men's interests in maintaining control over women continued to shape public policy after the transition to black majority rule in 1980 partly because a patriarchal mode of production continued to dominate subsistence agricultural production. In the current historical context patriarchal interests pose a major obstacle to socialist reforms and threaten the prospects for economic growth.[1]

Tributary Patriarchy in Precolonial Zimbabwe

Recent Marxist scholarship on Africa persuasively argues that few precolonial societies fit the idealized mold of "primitive communist modes of production" (Katz 1980). Rey (1973) and Meillassoux (1981) emphasize relations between the generations and define what they term "lineage" and "domestic" modes of production. Amin (1972; 1976) stresses the role of military power and relations between different ethnic groups in "tributary" modes of production. However welcome, these new efforts to analyze conflict and struggle within precolonial African societies remain somewhat incomplete. The terms "lineage" and "domestic" modes of production betray a reluctance to name their primary beneficiaries—elder men, or patriarchs. Neither Meillassoux nor Rey fully explore the exploitation of women. And the possible articulation between a mode of production based on the household and tributary relations between distinct ethnic groups remains largely unexamined.

Zimbabwe's historical record offers an opportunity to explore these new theoretical issues. Ethnographic data suggest that patriarchal relations were the primary relations of production among the Shona and were important among the Ndebele as well. Less is known of the tributary relations the militarily dominant Ndebele imposed upon the Shona, but they clearly complemented and in some ways built upon the patriarchal tradition. The Shona's patriarchal mode of production could and clearly did stand on its own; the tributary relations central to Ndebele society probably could not have existed independently of patriarchal relations, but they clearly exerted an important influence on the social formation as a whole.

On one point, virtually all analyses of Shona and Ndebele society concur: the household was the basic unit of production. The Shona were primarily engaged in subsistence agriculture (Beach 1977). The Ndebele had more cattle, and engaged in more trade, but also engaged in agriculture; some agricultural labor was performed by Shona captives (Cobbing 1974; Hughes and van Velsen 1955). In both groups, children began to provide labor at an early age, indicating a close relationship between reproduction and production. Yet Marxists have generally ignored relations within households and families. Arrighi's still classic description of the initial impact of colonization in Zimbabwe casually asserts that in the precolonial period "every adult member was entitled to land (which was abundant) in amounts sufficient for his and his family's subsistence" (1973, p. 223). Apparently not considered full adults, women had no independent access to land. Beach's well known essay on the Shona economy alludes to women in one sentence, noting only that they were legally classed as minors and "received little more than the food for which they worked" (1977, p. 55).

Not surprisingly, recent research shows that elder males controlled the means of production, land and cattle, among both the Shona and the Ndebele. They also controlled the means of reproduction: marriage rules and traditional law gave them considerable control over women and their reproductive capacities. Young men submitted to patriarchal control over part of their life cycle; women never escaped it. Elder men appear to have extracted a surplus from women both directly and by means of children (Muchena 1979; Weinrich 1979; England 1982).

Any attempt to analyze a history largely reconstructed after the fact by colonial administrators must be tempered by a certain skepticism. Nevertheless, a variety of sources offer a consistent account of patriarchal property rights. Among the Ndebele, in general children could not independently own property while their father was alive (Child 1968, p. 36). Hughes and van Velsen comment that while there were many qualifications in practice a Ndebele father theoretically controlled his son's economic activities (1955).

A Shona man was normally given his field by his father with the consent of the village headman only after his marriage (Bourdillon 1976, p. 43). Men became adults only upon marriage. Among both the Shona and the Ndebele, young men could marry only if their fathers provided them with a brideprice (*roora* or *lobola*) that was often "paid for" by a sister's brideprice. Among both peoples, nearly all marriages were arranged by parents (Hughes and van Velsen 1955, p. 99; Kuper 1955, p. 21).

Elder male control over children enhanced the economic benefits children provided parents, especially fathers. Brideprice was paid to a woman's father or brother (seldom to her mother) partially to compensate him for the loss of her labor but also in consideration of her reproductive capacities. The value of these transfers in labor terms is suggested by the fact that a man who could not afford a brideprice of cattle offered his services to his father-in-law for a specified period of time, normally ten years (Chigwedere 1982). Beach suggests that some men were forced to "sell themselves" to their father-in-law for life (1977, p. 55).

Children were valued not only as a source of male immortality, but also as a source of labor. Women were encouraged to have at least eight (Mutambwira 1979, p. 99). The economic importance of children is highlighted by their virtual equivalence with the value of cattle. Explaining the brideprice practices of both the Shona and the Ndebele, Chigwedere explains "if a man paid six head of cattle for a bride but the bride failed to give him children because she was infertile, he was given another bride (very likely his wife's sister). Alternatively if he married and paid six head of cattle but divorced after three children, the father-in-law refunded to him three head of cattle" (1982).

The economic advantages of high fertility provided strong incentives for strict control over women's reproductive capacities. A woman's consent to marriage was not required and many young girls were promised at an early age in return for a portion of the bridewealth payment. Once married a woman's claim or use-right to her husband's land could be revoked by divorce, and men could divorce women for a variety of reasons, including witchcraft. Among the Ndebele, a man could drive away his wife for any cause he deemed sufficient (Child 1968, p. 39).

Women who did not bear children were particularly vulnerable, apparently more so among the Shona than among the Ndebele, who did not necessarily consider it a woman's fault. Nonetheless, a barren woman was usually divorced unless her sister could be made available as a second wife. Voluntary childlessness was not tolerated. A man had a right to divorce his wife if she investigated or used birth control measures without his knowledge (Mutambwira 1979, p. 99).

Early European courts held that in Native Law "a man may always divorce a woman but a woman may never divorce a man" (Hughes, van

Velson 1955, p. 99). But among both the Shona and the Ndebele, a wife could dissolve a marriage only if her father or a guardian (such as an elder brother) gave his consent and returned the bridewealth. In the event of separation or divorce, fathers normally retained custody rights over children. Widows had no economic rights. A husband's possessions and his use-rights reverted directly to his male children, and in their absence, to his nearest blood relative (May 1983, pp. 65–69). Shona widows had considerable freedom in choosing other husbands, provided they were kinsmen of the deceased (Kuper 1955, p. 27).

Women in Shona society were considered legal minors. They required the sponsorship of a father, brother, or husband to participate in the traditional court procedures. The types of property a woman could own were strictly limited to small gifts in reward for motherhood and certain types of produce of the hands on *mavoko* property (May 1983, p. 65; Kuper 1955, p. 27). A Native Commissioner for Makoni observed in 1903 that "No woman can inherit property. She herself is property" (England 1982, p. 13).

Were these inequalities of access to the means of subsistence and production accompanied by exploitation within the family? Weinrich argues strongly that Shona men did indeed appropriate a surplus produced by women (1979). England pointedly suggests that the exploitation of women parallels that of peasants in Medieval Europe: "Each unit of mother and child in the polygamous Shona household had its own field and granary which maintained the male head, the mother and her children. However, the male head had exclusive rights to the produce of one field which served as insurance against the depletion of his wives' granaries . . . . The labour which women gave on this field can be readily identified as 'surplus' in the same sense that medieval peasants had obligatory services to perform on the lords' manorial lands" (1982, p. 12).

A definition of surplus based solely on women's agricultural labor is almost surely an underestimation, since men benefited disproportionately from women's childrearing activities. Some measure of labor time and intensity (including childcare time), relative to the size of the consumption bundle, would provide a better measure of exploitation (Folbre 1982; see also Henn, this volume). While there is no scope for measurement, this consideration strengthens the claim that women performed surplus labor that was partly appropriated by men.

The importance of exploitation within a patriarchal mode of production does not imply that it was identical to exploitation within other modes. The very kinship relations that provided a patriarch's labor force limited his control over it. While he could divorce a wife relatively easily, the brideprice requirement made it expensive for him to do so. Sons and daughters who disobeyed their father could be chased out of the family, but it took a rather long time to replace them.

Wives could and probably did resist patriarchal control. Hughes and van Velsen point out that "if a wife felt ill-used she could refuse to cook for her husband, and the other wives would frequently take her part and also refuse to cook for him" (1955, p. 94). Women were not merely engaged in material production, and the quality of the sexual and maternal services they provided surely suffered when they were abused. On the other hand, girls were socialized to their role from an early age, and patriarchal exploitation was a reliable, if not particularly lucrative source of surplus for men. The intensity of female socialization, and the pressures encouraging high fertility, no doubt impeded female rebellion.

These patriarchal features of precolonial Shona and Ndebele societies are similar to those of precolonial Cameroonian society (Henn 1978). They also conform to the description of other patriarchal societies largely un-articulated with capitalism, where patriarchal control over children provides strong economic incentives for pronatalist forms of control over women's reproductive capacities (Caldwell 1982; Stacey 1983).

But precolonial Zimbabwean society cannot be described simply in terms of a patriarchal mode of production, because relations between the Shona and Ndebele peoples as a whole were also important. The Shona's powerful centralized state disintegrated before the early 19th century, leaving them vulnerable to military domination by the Ndebele who had moved northward away from the Zulu. While British colonizers tended to exaggerate the extent of Ndebele predation, using it to justify establishment of colonial control, it is significant that it took a patriarchal form.

The tribute that the Ndebele appropriated from the Shona consisted primarily of women and children. A 19th century observer noted that there was nothing the Shona "deplored so much as their children being taken away from them just at a time when they become useful to their parents" (Cobbing 1974, p. 638). Shona captives belonged to the lowest caste among the Ndebele, but do not seem to have been treated any differently than Ndebele women and children from outside the royal family. They simply fulfilled their traditional role under a different set of patriarchs. For instance, Cobbing writes that a mature Ndebele warrior's family "grew with the addition of captives taken in war, distributed to him by the King or by the chief and distributed by him amongst his wives. In an analogous way a brave man could become wealthy in cattle" (Cobbing 1974, p. 630). A chain of authority linked the Ndebele individual, the kraalhead, the sectional headmen, the provincial chief, and ultimately the man known as the Big Father (Weinrich 1982, p. 37).

Patriarchal and tributary relations were not merely juxtaposed—they were interrelated, or articulated, in a way best conveyed by the term patriarchal tributary social formation. Sacks (1982) suggests a definite link between sexual inequality and the rise of state formations. Peoples such as the Mbuti of Zaire, with far less patriarchal relations than the Ndebele, were less inclined to pursue military conquest. In the Zimbabwean case, tributary social relations were premised on patriarchal relations, but not vice versa. Patriarchal relations were characteristic of the subjugated, as well as the dominant ethnic group. Patriarchal tribute of women, children and cattle became the accepted standard of military domination.

Perhaps as a result, the Shona may have anticipated the possibility of peaceful coexistence with the soldiers and settlers Cecil Rhodes led into their country in 1890. But six years later, realizing that the Rhodesians demanded adult male labor power and land rather than merely a tribute of women, children and cattle, they joined with their traditional Ndebele enemies in the first Chimurenga, or liberation war, in 1896. African military defeat signalled the beginning of a transition to a new social formation in which patriarchal relations of production continued to play a central role.

Patriarchal Capitalism, 1896–1940

The early colonial period in Zimbabwe is usually described entirely as a conflict between colonizers and colonized, with little or no consideration of patriarchal dynamics. True, the men who called themselves Rhodesians appropriated the best land, relegated the African population to native reserves, and took advantage of cheap labor to provide themselves with a standard of living better than most Europeans. But recent studies show that the native reserves system did not always work to the immediate benefit of white Rhodesians, and it affected African men and women in very different ways. White Rhodesians proved more than willing to reinforce at least some aspects of traditional patriarchal control, partly because such policies contributed to the consolidation of colonial control.[2]

The British South Africa Company, which ran the initial government in Rhodesia, recognized the need to utilize local structures of authority. As a means to this end, official government policy was to observe and enforce all aspects of customary law except those "contrary to natural law." With the exception of the outlawing of witchcraft, the distinction between laws that were "natural" and laws that were "contrary" pertained almost entirely to what settlers of English origin deemed the appropriate exercise of patriarchal power.

Initially, their policies were somewhat disruptive. They prohibited the pledging of young girls in infancy, declaring it illegal to give a woman in marriage without her own consent. The degree of enforcement of this provision is questionable, but its existence assisted the activities of Catholic and Protestant missionaries, whose first converts were young women seeking refuge from forced marriages (Ranger 1982, p. 8).

Tribal elders protested vociferously. Many of the civil cases the district commissioners dealt with after 1890 concerned the disobedience of daughters. Young women who resisted their father's choice of a husband often resisted their husbands as well, and the Native Commissioner from the Hartley District in Mashonaland stated in 1903 that the chief offense among Africans of his area was running away with other men's wives. In 1909 the Super-intendent of Natives at Gwelo expressed the same sentiment (Mittlebeeler 1976, p. 24). In Makoni district fathers and husbands insisted that colonial authorities should help them regain control over women (Ranger 1982, p. 10). The authorities proceeded to do so, partly as a concession, and partly because they worried that the "emancipation of women" had truly gotten out of hand.

The original prohibitions were not rescinded, but forms of patriarchal control over adult women were reinforced. The Rhodesians had never declared polygyny illegal, and while they made registered Christian marriages available, they did not require them. In 1916 they took steps to "strengthen" African marriage. A new penal enactment stipulated a fine not exceeding 100 pounds, or imprisonment with hard labor for a period not exceeding one year against both the African man and woman committing adultery (Mittlebeeler 1976, p. 124). Though superficially symmetrical in its impact, this law affected women very differently than men. Since polygyny was not outlawed, and because traditional unregistered marriages were quite flexible, a man could only be convicted of adultery if his sexual partner was another man's wife. Any woman who had ever been married, however, was vulnerable to prosecution.

Ranger writes convincingly of the enthusiasm with which white Rhodesians worked together with tribal elders to solve the "woman problem." In urging the passage of the adultery law for Africans the Native Commissioners of Rusape and Umtale explained that the "Native Marriages Ordinance and settled government has conferred on native women a freedom previously unknown to them. This freedom has in many cases provided a cloak for license" (Ranger 1982, p. 13). Even the missionaries repented, changing their educational policies to emphasize home-making rather than employment skills for young women.

In 1917 the Missionary Women's Conference acknowledged that they had preached too much liberty to women and decided to recommend the preaching of Paul's injunction, "Let the wives be subject to their husbands." However, the Apostolic Church and other offshoots from the major religious denominations continued to promote female emancipation and received a number of defectors from the Methodist and Catholic missions. Shortly thereafter they lost the support of the Native Affairs Administration (Ranger 1982, p. 13).

Collaboration between the Native Affairs Administration and tribal patriarchs set limits, albeit not completely successful ones, on women's participation in the capitalist sector. In 1927 Native Commissioner Holland observed that "every man is opposed to a native woman leaving her kraal because her productive labor was his asset" (England 1982, p. 32). Commissioner Janning reported that the chiefs in his area opposed female emancipation and that "chiefs wanted a pass system for women introduced to restrict the number of women who 'escaped' by taking trains" (England 1982, p. 52). In the 1940s and 1950s the number of African women in the cities increased considerably and in 1946 the Native Commissioner of Marandellas expressed his discouragement "that a large number of women make their way to towns in spite of every obstacle raised to prevent them" (England 1982, p. 73).

While urban employment opportunities and new forms of transportation destabilized traditional patriarchal relations, white Rhodesian employment policies bolstered them. Women made few inroads into wage employment, despite their willingness to work for even lower wages than men. The 1949 legislation that set a minimum wage and stipulated labor conditions for African men mentioned African women not at all. Even as late as 1969 (the first year for which data are available) only 5.6 percent of women between the ages of fifteen and sixty-four were employed in wage labor, and over 50 percent of these were engaged in the lowest paying agricultural work (1969 Census, Tables 10, 77).

One might suppose that an ample supply of male labor made restrictions on women's employment virtually costless to white employers, but most historians agree that labor remained relatively scarce until at least the 1940s. The native reserves set aside for the Africans, despite their poor soil quality and lack of rainfall, proved more of a refuge than anticipated. Despite the imposition of a hut tax, and later, a poll tax, designed to force African men into wage labor, white owned mines and farms suffered from a shortage of willing labourers (Kay 1982, p. 85). Extensive recruitment was necessary to mobilize labor from Malawi, Zambia, and Mozambique, and in 1911, of the 30.4 percent of all working age males that were in wage employment, 69 percent were foreign born. The foreign born comprised more than one half the wage labor force until 1950 (Johnson 1964B, p. 169).

Although we do not know the age distribution of Zimbabwean wage workers, it is likely that many were young men who had not yet received land rights within the reserves or who were dissatisfied with their land allotment. Young men had a special incentive to seek employment, even at exploitatively low wages: they could anticipate earning enough money to pay their own brideprice and thus to marry without their fathers' consent (Chigewedere 1982, p. 41). Some observers note that elopement became increasingly common (Weinrich 1982, p. 61). In the 1930s the Native Commissioner for Bulalima-Mangwe deplored the effect of urban migration on the older generation: "The result is that the old men and womenfolk have to bear the burden and see to the homes and stock, for impatient of parental control the boys . . . readily desert the flocks in the field" (England 1982, p. 33).

Yet the patriarchs still maintained substantial authority. A father could legally claim the earnings of his unmarried children and exercise control over any property acquired by them. More importantly, extremely low wages and insecurity of employment in the capitalist sector forced young men to maintain their ties to the rural areas. Those who failed to remit a portion of their earnings to their fathers might find themselves unwelcome at home when there was literally no other place to go. Even those who expected to live primarily in urban areas wanted land where their wife and children might live.

For a while, at least, the reserves offered a viable alternative to wage labor. The expansion in cultivated African land area actually exceeded population growth until the mid 1940s, primarily due to a rapid increase in plows from 3,402 in 1911 to 108,410 in 1941 (Johnson 1964B, p. 175). The Ndebele cattle herds grew rapidly after the initial trauma of conquest and infectious disease. The Shona were initially able to sell much of their agricultural output. A battery of new laws drastically restricted the ability of African farmers to compete with Europeans (Palmer 1977, p. 229). But as late as 1946 only 39 percent of the African population lived outside the Tribal Trust Lands, and a much smaller percentage anticipated living permanently outside (Johnson 1964B, p. 169).

These observations suggest that the explanation of Native Reserves systems developed by Wolpe (1980) is oversimplistic. He argues that such systems primarily reflected employer's interests in minimizing labor costs. But capitalist logic alone does not explain the native reserve system in Zimbabwe, which probably contributed to labor shortages and increased labor costs until at least the 1940s. Patriarchal logic was also at work. Restrictions on women's participation in the modern sector, favored by many African men, purchased a form of political stability that was particularly costly to women. Men were able to leave the reserves to work for wages without sacrificing their control over production there. If the women and children who remained behind lowered the cost of labor power, they did so to the benefit of African men as well as white employers. The benefits that older men enjoyed as a result served as small, but not insignificant padding against the burden of colonialism.

Immiseration and Resistance: 1940–1980

However expedient the native reserve system proved for a time, its days were inevitably numbered. As early as 1941, observers began to comment on mounting population densities. Virtually all explanations of the economic impetus behind the guerrilla war of the 1970s in Zimbabwe stress that population pressure led to a steady deterioration in the rural African standard of living (Riddell 1978; Seidman, Martin, and Johnson 1982). But, contrary to conventional accounts, population growth was not a natural, or exogenous event. The persistence of strong patriarchal power contributed to high birth rates in rural areas. Furthermore, the structure of the patriarchal mode of production shifted much of the economic stress onto a younger generation and thereby influenced the character of political struggle.

Between 1940, the year when population pressure on the land was officially recognized, and 1962, the first year the Rhodesians actually conducted a complete census, the African population doubled in size. Between 1962 and 1980 they doubled again. The growth rate of the African population steadily accelerated, from an average of 2.3 percent per annum between 1901 and 1920, to 2.7 percent between 1920 and 1940, to 3.3 percent from 1940–1970 (1969 Census of Population, Monthly Digest of Statistics, March 1983). The immediate cause of this acceleration was an increase in African life expectancy, despite a stagnating and ultimately declining rural standard of living.

White Rhodesians recognized, from the outset, that they had a vested interest in improving the health of the African population. In the early 1900s high mortality in the mines exacerbated labor shortages and led to the imposition of some basic health regulations on individual firms (Van Onselen 1976, pp. 48–60; Gelfand 1976). The risk of contagion motivated substantial smallpox immunization campaigns. As a letter from the Medical Director of Salisbury to the Director of Native Development explained in 1930, "as we want to have a healthy white nation we have to tackle infectious diseases in the native. The native is the reservoir of these infectious diseases" (Riddell, Saunders, Gilmurray 1979, p. 33).

Health care available to Africans remained far inferior to that for white Rhodesians. Malnutrition was endemic. But immunization, quarantine control, and the dissemination of modern antibiotics through a minimal rural health care system proved relatively inexpensive and somewhat effective. By 1950 Zimbabwe had the lowest overall mortality rate in Southern Africa, and life expectancy continued to improve between 1950 and 1980 (UN 1981).

But mortality decline alone did not account for rapid population growth. Fertility rates remained extremely high. Women continued to marry young and to bear large numbers of children within marriage. Desired family size remained high, partly because children continued to make important contributions to household production. Johnson's study of the Chiweshe Reserve in the 1960s showed that children accounted for over 25 percent of agricultural labor (1964A, p. 18). Women, who were largely restricted to subsistence agricultural production correctly perceived that the opportunity cost of time spent in childrearing was relatively low. In contrast, the cost to men was even lower, since women and children normally paid their own subsistence costs even if they did not generate a surplus (Folbre 1983). Rhodesian policy reinforced patriarchal incentives to high fertility, not only by restricting African women's opportunities, but also by strictly limiting educational opportunities for African children.

The government was initially more concerned about the effect of African population growth on land than on social conditions. Overgrazing and erosion in some rural areas (overstated by many colonial administrators) motivated the first official effort to actually modify the social relations of production in the native reserves. The 1941 Natural Resources Act was designed to limit stock to the calculated carrying capacity of the land and to allocate individual rights in arable lands. While it was not applied successfully everywhere, it established a legal basis for individual male tenure and grazing rights.

During the boom years between 1946 and 1951, officials even spoke of eliminating the native reserve system. Interest in creating a permanent class of wage workers, as well as apprehensions about the fragmentation and misuse of farm land led to the African Land Husbandry Act of 1951, which reinforced the formal male privatization set in motion ten years earlier and regulated the actual size of holdings by establishing both a minimum and a maximum (Kay 1980, p. 90).

From the point of view of political strategy the African Land Husbandry Act of 1951 was an unmitigated disaster. The only way to formally designate landowners was to formally designate the landless. It quickly became apparent that there was simply not enough land to go around, and implementation of the act proceeded very slowly. In 1962 the Secretary for Internal Affairs conceded that "alternative outlets have not kept pace with the increase in population" (Kay 1980, p. 90). Rhodesians became worried enough to conduct their first census of the African population in 1962. To their chagrin, they discovered that the African population was 20 percent greater than they had estimated. Fear of being swamped by the growing black population contributed to the defeat of the reformist government and the ascendance of the Rhodesian Front, led by Ian Smith, which unilaterally declared independence from Great Britain in 1965.

The Smith government sought to solve the problem of landlessness by political fiat. The Tribal Trust Land Act of 1965 nullified the African Land

Husbandry Act of 1951 and returned responsibility for allocating land to the Africans, without strictures on parcel size. Smith took pains to cultivate the loyalty of tribal patriarchs, and sometimes succeeded. Needless to say, rural discontent related to landlessness did not abate. It intensified.

Rural discontent was mediated by rural demographics. By 1962, 82 percent of the population of the Tribal Trust Lands consisted of women and children. Almost 70 percent of all African children under the age of 14 lived there (1962 Census). Little is known of the size or frequency of the remittances that husbands and fathers made, but they were not sufficient to maintain incomes in the reserves at a level approaching that of urban or even commercial agricultural areas. Indeed Sutcliffe argues that the urban/rural income differential increased drastically after 1950 (1971).

The growing scarcity of land had a particularly severe impact on women and young men: women, because they were especially dependent on agricultural production; young men because their chances of ever owning land were decreasing. Weinrich's study of 902 men in two Tribal Trust land areas showed that 5.5 percent of men over forty-five were landless whereas 29.1 percent of men 30–44 and an overwhelming 80.7 percent of men 15–29 were landless (1975). In 1969 nearly three fourths of all the unemployed in Rhodesia were under 30 years old (Clarke 1977, p. 30).

Young men and, to a lesser extent, young women began to assume considerable political leadership. The organization which spoke out most forcefully against minor reforms and in favor of majority rule was the City Youth League founded by James Chikerema, George Nyandaro and Edson Sithole (Seidman, Martin, and Johnson 1982). When the struggle moved into its military phase in the mid-1960s, the rural combatants, both male and female, were conspicious for their youth. They called themselves and were referred to by Rhodesian Security Forces, as "the boys." Some persuaded their parents to join the cause. When Mzanyan Ndlovu, headman of Gwana, was asked why he supported the guerillas, he said, "because they are our sons and daughters" (Frederickse 1982, p. 81).

But the white Rhodesians appealed to their official patriarchs. They lined up a large group of tribal chiefs willing to support Unilaterally Declared Independence, and when the British Prime Minister refused to meet with them on the grounds that they were unrepresentative, the chiefs protested that representation was not an issue. They pointed out, accurately if unsuccessfully, that they had inherited their position just as the Queen of England had. During the war Information Minister P.K. van der Byl hoped to combat the guerillas by "giving every chief total control of his own 'fiefdom' to prevent black people from stepping out of line and getting subversive" (Frederickse 1982, p. 81). This policy eventually backfired and discredited chiefly collaborators, because "the boys" were willing to promise something the patriarchs could not—land.

By 1979 the Chimurenga or Liberation forces had successfully established their political credibility and their military strength, and agreed to negotiate a settlement largely because neighboring states pressured them to end a war that was disrupting the regional economy. An agreement made at Lancaster House laid the foundation for the election that brought Robert Mugabe and the Zimbabwe African National Union (ZANU) to power.

Many complex factors contributed to the success of this struggle, but it was impelled by a confluence of economic and demographic pressures. Growing poverty, paradoxically combined with decreased child mortality, might have encouraged African women to lower their fertility rates, but the persistence of strong patriarchal relations within the reserves made it difficult to do so. Many elder African men might have been reluctant to risk revolution, but their sons and daughters felt the full force of both colonial and patriarchal subordination, and rebelled. In many respects, black majority rule represented a decisive break with the past. Yet it also, perhaps inevitably, preserved much of the legacy of patriarchal capitalism.

Patriarchy and Black Majority Rule

The idealized vision of precolonial Africa as a collection of primitive communist modes of production has tempted some to argue that a transition to socialism would come easier there than elsewhere, but that is not so. In Zimbabwe, Robert Mugabe voiced a strong moral committment to socialism, and resettled over 21,000 people on previously white owned land. But the collective farms in his original land reform plans never materialized. Nor have any major industries been nationalized.

The lack of structural economic change can be partially explained by fear of the type of capital flight that disabled the Mozambican economy as well as the military threat posed by South Africa (Yates 1982). Ranger offers more of a class analysis, making a strong case for the importance of a "consistent peasant political ideology and programme" and noting that peasants as a class are hardly ever committed to "collective agricultural production systems" (Ranger 1985, p. 289). All these explanations overlook the possibility that another source of political inertia lay in male peasants' reluctance to relinquish their own patriarchal privileges.

A brief consideration of events in Zimbabwe since 1980 lends some credibility to this hypothesis. In a pattern reminiscent of recent Chinese history (Stacey 1983), Zimbabwean policies have weakened the power of fathers over children but reinforced the power of husbands over wives. The most significant legal reform instituted by the Mugabe government was the 1982 Legal Age of Majority act. Both African men and women were granted legal majority upon reaching age 18, allowing them to vote and to marry legally without their parent's permission. Thus for the first time African women had the right to enter into legal contracts.

When the prime minister met with over 200 traditional chiefs in March of 1985 to discuss his policies, this legislation was the biggest bone of contention. The chiefs worried that parents could no longer control children over eighteen. Senator Ndweni elaborated: "How can children be allowed to marry without consulting their parents? We cannot allow children to do whatever they want. We are destroying the whole nation" (Ranger 1985). Mugabe diplomatically reassured the chiefs that the law had been intended primarily to extend the franchise.

Widespread opposition to even this small reform has inhibited changes in other aspects of family law. And while many women within the Government have openly committed themselves to feminist goals, the dominant political party remains influenced by the orthodox Marxian notion that women's subordination will automatically recede as women enter the wage labor force (Roberts 1984). As a result, many aspects of traditional African law that disadvantage women have remained in place (Rule 1986; Jacobs and Howard 1985; Seidman 1984).

Sex discrimination is virtually the only form of discrimination that is not outlawed in the new Zimbabwean constitution. Women have no legal guarantees of joint ownership, inheritance from husbands, or even control over earnings. Asserting the importance of national tradition, Chigwedere explains that a law such as "the wife has complete autonomy over her salary will in practice cause upheavals in families" (1982).

Of course, access to land for women would cause even greater turmoil. Agricultural collectives and cooperatives would not necessarily offer women more control than they now have, but could certainly loosen patriarchal control. The only significant agrarian reform policy in Zimbabwe is the resettlement program, based on government purchases of land from white farmers. Only settlers who are married or widowed with dependents are eligible. While some widows have received land, women who are on their own, deserted or divorced are largely excluded.

Cheater argues that male farmers in some areas see women as their agricultural labor force and are therefore largely opposed to a proposed law that wives should automatically inherit their husband's land (1981). A survey carried out by the Zimbabwean Women's Bureau documented widespread dissatisfaction among rural women. Women were nearly unanimous in their desire for land rights and they used strong language to describe men's behavior in controlling land, calling them "exploiters" and "bloodsuckers" (Muchena 1979).

Unless patriarchal control over women and young children diminishes, strong economic incentives for high fertility will remain in place. In December of 1985 a newspaper article on the Mukohwe Valley made the following observation: "Talk of family planning programmes has been heard in the Valley but most people there have always held and still believe that a large labour force of wives and children in the field means big food production. Others believe that it is a custom which should be sustained" (Herald, Dec. 3, 1985).

Family planning assistance is available. But women's ability to take advantage of it is limited by male disapproval. A report published by the Ministry of Community Development and Women's Affairs in 1982 showed that Zimbabwean men overwhelmingly oppose contraception on the grounds that it makes wives unfaithful (1982, p. 74). Jacobs notes two topics consistently brought up at Women's Club meetings in resettlement areas: (1) the need for safe, effective child spacing methods which would not be visible to men and (2) the right to refuse sex. Refusal of sex with husbands is one of the chief reasons for wife beating (1985, p. 7). A more recent survey suggests that family planning is gaining more acceptance (Boohene 1986), but abortion in Zimbabwe remains illegal, even when the mother's life is endangered (Seidman 1984, p. 435).

One major government policy, the massive expansion of primary school enrollments, may help diminish fertility in the long run. On the other hand, the new commitment to rural health may significantly lower mortality rates and therefore increase population growth rates. Even at its recent historical growth rate of 3.5 percent per year, the African population of Zimbabwe will reach 9.5 million in 1990, outstripping even the most optimistic plans for agricultural resettlement. Landlessness is bound to increase, and women are likely to suffer most.

Conclusion

A better understanding of the patriarchal past could mean a less patriarchal future. The concept of a patriarchal social formation poses a constructive challenge to the modern Marxist tendency to marginalize feminist theory. It highlights the complementarity between patriarchal and tributary relations in precolonial Zimbabwe, and helps explain policies towards women that bolstered the native reserve system during the early colonial period. It offers some insight into the persistence of high fertility rates among African women and the role of youth in the war of liberation. Finally, it identifies an important source of resistance to agricultural collectivization. On a more general level, the concept of a patriarchal social formation incorporates inequalities based on gender as well as nation, race, and class into a new more accurate map of Zimbabwe's political economy. And while many blank spaces need to be filled in, this map provides a far better guide to the future than those which shove inequalities between men and women to one side.

Notes

1. Some of these individual points, as well as most of the evidence supporting them, are drawn from a large body of recent historical research in Zimbabwe that has been informed by feminist concerns. Since much of this work remains unpublished, I would like to acknowledge my debt to it at the very outset, and emphasize that my primary aim here is to show how the concept of a patriarchal social formation can weave these historical findings into a unified reinterpretation of Zimbabwean history that places gender on the same level of theoretical importance as race and class.

2. An exploration of the patriarchal dimensions of white Rhodesian society is beyond the scope of this paper, but it is important to note that the benefit the white population derived from its expropriation of African land and labor diminished the impact of both class and gender inequalities within white society.

A 1924 Guide to Rhodesia stated, "there is room for almost any type of settler who is willing to work intelligently, except the unskilled labour classes, since all unskilled labour is performed by natives" (Kay 1980, p. 97). The Guide could easily have added "women who choose to accompany their pioneer husbands will find ample opportunity for leisure as natives are easily trained to assume household and childcare responsibilities." Male domestic servants could be hired for a pittance, and were widely utilized.

Early Rhodesian restrictions on white women's legal and economic rights mirrored those of Victorian women, but white Rhodesian women, unlike their English counterparts, never became particularly exercized over feminist issues. One of the few official manifestations of debate over women's rights during the entire colonial period was a mildly worded "Report of the Commission of Inquiry into the Inequalities or Disabilities between Men and Women" presented to the Legislative Assembly in 1956. But while white women never posed problems for Rhodesian policy, African women were a subject of major concern.

References

Amin, S. (1972) "Underdevelopment and dependency in black Africa-origins and contemporary forms," *Journal of Modern African Studies* 10:4, 503–524.

Amin, S. (1976) *Unequal Development*. New York: Monthly Review Press.

Arrighi, G. (1973) "Labor supplies in historical perspective: a study of the proletarianization of the African peasantry in Rhodesia," in Giovanni Arrighi and John S. Saul, eds., *Essays on the Political Economy of Africa*. New York: Monthly Review press.

Banaji, J. (1977) "Modes of production in a materialist conception of history," *Capital and Class* 3: 1–44.

Beach, D. (1974) "Ndebele raiders and Shona power," *Journal of African history* 15, 4: 633–651.

Beach, D. (1977) "The Shona economy: branches of production," in R. Palmer and N. Parsons, eds., *The Roots of Rural Poverty in Central and Southern Africa*. Berkeley: University of California Press.

Boohene, E.S. (1986) "Summary of the results of the 1984 Zimbabwe Reproductive Health Survey." *Zimbabwe Science News* 20, 7/8: 83–88.

Boserup, E. (1970) *Women's Role in Economic Development.* London: Allen and Unwin.

Bourdillon, M. (1976) *The Shona Peoples.* Harare: Mambo Press.

Caldwell, J. (1982) *The Theory of Fertility Decline.* New York: Academic Press.

Cheater, A. (1981) "Women and their participation in commercial agricultural production: the case of medium-scale freehold in Zimbabwe," *Development and Change* 12: 349–77.

Chigwedere, A. (1982) *Lobola.* Harare: Books for Africa.

Child, H. (1968) *The amaNdebele.* Harare: Ministry of Internal Affairs, Rhodesian Government.

Clarke, D.G. (1977) *Unemployment and Economic Structure in Rhodesia.* Harare: Mambo Press.

Cliffe, L. (1982) "Class formation as an articulation process: East African cases," in Hamza Alavi and Teodor Shanin, eds., *Introduction to the Sociology of Developing Societies.* New York: Monthly Review Press.

Cobbing, J. (1974) "The evolution of Ndebele Amabutho," *Journal of African History* 15, 4: 607–631.

England, K. (1982) "A Political economy of black female labor in Zimbabwe, 1900–1980." B.A. thesis, Honors School of History, University of Manchester.

Folbre, N. (1986) "A patriarchal mode of production," in Randy Albeda and Christopher Gunn, eds., *New Perspectives in Political Economy.* New York: M.E. Sharpe.

————. (1983) "Of patriarchy born: the political economy of fertility decisions," *Feminist Studies* 9:2 (Summer): 261–284.

————. (1982) "Exploitation comes home: a critique of the Marxian theory of family labour," *Cambridge Journal of Economics* 6: 317–329.

Foster-Carter, A. (1978) "The modes of production controversy," *New Left Review* 107: 47–77.

Frederickse, J. (1982) *None But Ourselves.* Harare: Zimbabwe Publishing House.

Gelfand, M. (1976) *A Service to the Sick: A History of the Health Services for Africans in Southern Rhodesia, 1890–1953.* Harare: Mambo Press.

Henn, J. (1978) Peasants, Workers and Capital: The Political Economy of Labor and Incomes in Cameroon. Ph.D. Dissertation, Harvard University.

Hindess, P. and P. Hirst, (1975) *Pre-Capitalist Modes of Production.* Boston: Routledge and Kegan Paul.

Hughes, A.J.B. and J. van Velsen (1955) *The Ndebele.* London: International African Institute.

Jacobs, S. (1986) "Women in Zimbabwe: stated policy and state action," in H. Afshar, ed., *Women and the State: Studies from Africa and Asia.* London: Macmillan.

————. (1985) "Some issues in land resettlement in Zimbabwe: experieces of Shona women in resettlement schemes," Discussion Paper, Institute of Commonwealth Studies, Centre for African Studies, Univesity of London, Dec. 3.

————. (1984) "Women and land resettlement in Zimbabwe," *Review of African Political Economy* 27/28: 33–50.

Jacobs, S. and T. Howard. (1985) "Women in Zimbabwe: Stated Policy and State Action," unpublished manuscript, forthcoming in H. Afshar, ed., *Women and the State.*

Johnson, R.W.M. (1964a) "The labour economy of the Reserves," Occasional Paper 4, Department of Economics, University College of Rhodesia and Nyasaland.

———. (1964b) *African Agricultural Development in Southern Rhodesia: 1945–1960.* Palo Alto: Food Research Institute, Stanford University.

Katz, S. (1980) *Marxism, Africa, and Social Class: A Critique of Relevant Theories.* Centre for Developing Area Studies, Occasional Monograph Series, No. 14, McGill University, Montreal, Canada.

Kay, G. (1980) "Toward a population policy for Zimbabwe, Rhodesia," *African Affairs* 79, 314 (January).

———. (1982) "Population distribution in Zimbabwe," in John J. Clarke and Leszek A. Kosinki, eds., *Redistribution of Population in Africa.* London: Heinemann.

Kuper, H. (1955) *The Shona.* London: International African Institute.

Mamdani, M. (1976) *Politics and Class Formation in Uganda.* New York: Monthly Review Press.

May, J. (1983) *Zimbabwean Women in Customary and Colonial Law.* Harare: Mambo Press.

———. (1979) *African Women in Urban Employment: Factors Influencing Their Employment in Zimbabwe.* Harare: Mambo Press.

Meillasoux, C. (1981) *Maidens, Meal and Money: Capitalism and the Domestic Community.* Cambridge: Cambridge University Press.

Mittlebeeler, E. (1976) *African Custom and Western Law: The Development of the Rhodesian Criminal Law for Africans.* New York: Africana Publishing Company.

Mosley, P. (1983) *The Settler Economies: Studies in the Economic History of Kenya and Southern Rhodesia, 1900–1963.* Cambridge: Cambridge University Press.

Muchena, O. (1979) "The changing position of African women in rural Zimbabwe-Rhodesia," *Zimbabwe Journal of Economics* 1 (1): 44–61.

Mutambwira, J. (1979) "Traditional Shona concepts of family life," *Zimbabwe Journal of Economics* 1:2 (June).

Ong, B.N. (1985) "Women in the transition to socialism in sub-saharan Africa," in B. Munslow, ed., *Africa's Problem's in the Transition to Socialism.* London: Zed Press.

Palmer, R. (1977) "The agricultural history of Rhodesia," in R. Palmer and N. Parsons, eds., *The Roots of Rural Poverty in Central and Southern Africa.* Berkeley: University of California Press.

Poulantzas, N. (1973) *Political Power and Social Classes,* translated by Timothy O'Hagan. London: New Left Books.

Ranger, T. (1985) *Peasant Consciousness and Guerrilla War in Zimbabwe.* Berkeley: University of California Press.

———. 13th Review of the Zimbabwe Press. Unpublished manuscript, Department of History, University of Manchester.

———. (1982) "Women in the politics of Makoni District Zimbabwe, 1890–1980," Manuscript, Department of History, University of Manchester.

Report of the Commission of Enquiry into the Inequalities or Disabilities Between Men and Women. (1956) Legislative Assembly, Salisbury, Rhodesia. Zimbabwe Archives.

Rey, P. (1973) *Les Alliances de Classe*. Paris: Maspero.

Riddell, R.C. (1978) *The Land Problem in Rhodesia*. Harare: Mambo Press.

Riddell, R., D. Saunders and J. Gilmurray (1979) *The Struggle for Health*. Harare: Mambo Press.

Roberts, P. (1984) "Feminism in Africa, feminism and Africa," *Review of African Political Economy*, 27/28: 175–185.

Rule, S. (1986) "Zimbabwe still divided on rights for women," *New York Times*, March 27.

Sacks, K. (1982) *Sisters and Wives: The Past and Future of Sexual Inequality*. Chicago: University of Illinois.

Seidman, G. (1984) "Women in Zimbabwe: post-independence struggles," *Feminist Studies* 10,3.

Seidman, G., D. Martin and P. Johnson, (1982) *Zimbabwe: A New History*. Harare: Zimbabwe Publishing House.

Stacey, J. (1983) *Patriarchy and Socialist Revolution in China*. Berkeley: University of California Press.

Stitchter, S. (1982) *Migrant Labour in Kenya: Capitalism and African Response*. London: Longman.

Sutcliffe, R.B. (1971) "Stagnation and inequality in Rhodesia, 1946–68," *Bulletin of the Oxford University Institute of Economics and Statistics 33*, 1 (February).

Terray, E. (1972) "Historical materialism and segmentary lineage based societies," in E. Terray, *Marxism and Primitive Societies*. New York: Monthly Review Press.

United Nations. (1981) *World Population Prospects as Assessed in 1980*. New York.

Van Onselen, C. (1976) *Chibaro: African Mine Labour in Southern Rhodesia*. London: Pluto Press.

Weinrich, A.K.H. (1982) *African Marriage in Zimbabwe*. Harare: Mambo Press.

––––––. (1979) *Racial Discrimination and Women in Zimbabwe*. Geneva: Unesco.

––––––. (1975) *African Farmers in Rhodesia*.

Wolpe, H. (1980) "Capitalism and cheap labour-power in South Africa: from segregation to apartheid," in H. Wolpe, ed., *The Articulation of Modes of Production*. London: Routledge and Kegan Paul.

Yates, P. (1982) "The prospects of socialist transition in Zimbabwe," *Review of African Political Economy* 18 (May-August): 68–88.

4

Demographic Theories and Women's Reproductive Labor

Jane Vock

Introduction

The objectives of this paper are twofold: first to provide a discussion of the relevance of feminist theory to theories of demography, and vice versa; and second, to outline the problems which arise when theories of demography are not informed by feminist scholarship. While the first of these objectives will entail a more abstract theoretical focus, the second will be achieved through an examination of the work of Africanist demographers, Joel Gregory and Victor Piché, and John Caldwell, particularly their explanations of fertility patterns and behaviour.[1]

Gregory and Piché, as Marxist demographers, provide a refreshing challenge to the conventional wisdom and dominance of Malthusian and demographic transition theories.[2] From a feminist perspective, however, the present conceptualization of this alternative framework is problematic. Broadly stated, they fail to consider or incorporate feminist insights on gender relations and the subordination of women. This is a problem shared by most demographic theorists and is by no means confined to Marxist analyses of demographic phenomena such as that of Gregory and Piché. While conventional population theory has begun to include work on women, this usually focusses on issues such as women's labor force participation and fertility patterns and behavior. The linkages made between women's productive and reproductive roles rarely address women's *subordination*. This omission leads to a subsequent neglect of the implications or consequences

The author wishes to thank Rhoda Howard and Sharon Stichter for their helpful comments and suggestions, many of which have been incorporated into this paper.

that male dominance or patriarchy has for demographic patterns and behaviour.

While greater attention to the social relations of human reproduction is crucial for the development of feminist scholarship itself, this neglect on the part of Marxist demographers impedes a more complete understanding of demographic phenomena.[3] In the context, although Gregory and Piché's Marxist model of fertility yields more explanatory power than Caldwell's demographic transition theory, Caldwell provides a more perceptive treatment of the social relations of reproduction.

Feminist Theory and Women's Human Reproductive Labor

First we must clarify the concept of reproduction. In response to the loose manner in which this concept has been employed, Edholm, Harris and Young (1977) have made a useful analytical distinction between the types or kinds of reproduction: social reproduction; reproduction of the labor force; and biological/human reproduction. They correctly assert that the lack of clarity in this concept's usage and/or the conflation of these types of reproduction often leads to theoretical confusion and error. Of relevance to this discussion of demographic theories is the conflation of biological reproduction and reproduction of the labor force. While the two are inextricably connected, given that babies develop into children and later adults and workers, such a conflation leaves no conceptual space to consider how each may have its own unique determinants and dynamics. Our understanding of childbearing is reduced to nothing more than the reproduction of labor power. Furthermore, such a conflation obfuscates the physical costs of childbearing, which are obviously borne by women alone since childbearing is sex-specific. While men may absorb some of the costs of childrearing, primarily through wages or other contributions toward feeding and clothing the mother and child(ren), it is women who suffer the health costs and risks of death associated with pregnancy and childbearing. This of course is particularly important for women in underdeveloped countries, where such costs and risks are extremely high.[4] Given the confusion which can and often does arise over the concept of reproduction, it is necessary to make my own position explicit. I agree with Edholm, Harris and Young that human/biological reproduction needs to be analytically distinguished from other types of reproduction and in particular from the reproduction of the labor force. While childbearing is the reproduction of labor power, it is not simply or only this. Such a distinction raises the issue of fertility and childbirth to a status of more theoretical importance and relatedly enhances a deeper understanding of its dynamics and determinants as a separate labor process.

A further theoretical issue which has been highlighted by feminist scholars, and an issue most often ignored within theories of demography, is men's control of women's fertility. A clear link is postulated between the control of women's biological reproductive capacities and patriarchy: the social relations which govern human reproductive behaviour serve to reflect, instill or reinforce the subordination of women. To cite just a few examples of this assertion: "Patriarchal oppression of women in the family is crucially connected with the need to control their fertility and sexuality" (McDonough and Harrison 1979, p. 14); "The characteristic relation of human reproduction is patriarchy, that is control of women, especially of their sexuality and fertility, by men" (Mackintosh 1977, p. 122); "Patriarchal authority is based on male control over the women's capacity and over her person" (Rowbotham 1973, p. 117).

Clearly, whether patriarchy is defined as the rule of the father or the domination of women by men; whether it is seen as inseparable or autonomous from the social relations of production; or whether it is seen as materially and/or ideologically based, there is a consensus that the control of women's childbearing capacities is integral to the operation of patriarchy.

What is lacking, however, is theoretical and empirical work which attempts to probe further into the social processes of biological human reproduction and the laws of motion of patriarchy in this context. There is a need for greater reflection and sophistication regarding the specificity of these relations and the interconnectedness between patriarchy, fertility patterns and behaviour, and the oppression of women. In this regard, too many feminist writers merely proclaim the existence of men's control over women's fertility. Fertility may be, however, one area where males cannot successfully exercise complete domination. While agreeing that patriarchal control and authority generally dominate the social relations of reproduction, it is important to raise and confront certain questions surrounding this issue. Under what conditions are women expected to bear more or fewer children? Under what conditions are men and women (husband and wife) in congruence in relation to desired family size? Or conversely, when are gender differences evident in relation to both power in decision-making and desired family size? And finally, in what way, and under what conditions do men benefit from control over women's fertility?

Demographic theories and empirical examinations of demographic phenomena are clearly of utmost importance to the further development of feminist scholarship. They have the potential to provide insights into the material and/or ideological bases of patriarchal control of women's fertility. Feminist theory, in turn, can make a valuable contribution to theories of demographic processes and behaviour. It can be argued, in fact, that any demographic theory which excludes a discussion of intrahousehold dynamics, and which does not take into account unequal gender relations is, at best, incomplete, and at worst, distorted. In the following sections of this paper, some of these limitations and distortions will be addressed in greater detail through an outline and critique of the work of Gregory and Piché and of Caldwell.

Explanations of High Fertility in Contemporary African Societies

Within the Marxist framework, fertility is typically conceptualized as a response to the demand for labor. The production of babies is the production of labor power or human energy. This formulation includes the demand for labor inherent in a local production system and/or a demand imposed from outside (Franke 1918). It is in this context that we can place Gregory and Piché's analysis of fertility patterns and behaviour in contemporary African social formations.[5]

The explanation of African underdevelopment which emerged primarily from the French school of Marxist anthropology serves as the base for Gregory and Piché's model of fertility.[6] Stated tersely, the nature of peripheral capitalist economies is seen to be the assymetrical articulation of the capitalist (dominating) and domestic (dominated) modes of production (Gregory and Piché 1981, p. 22). The intrusion of capitalism/colonialism in Africa transformed the domestic mode such that it has lost both its autonomy and precapitalist/precolonial character. This noncapitalist domestic economy is no longer able to provide for its own sustenance (an aspect of the "freeing of labor"), but is, at the same time, necessary, because the wages obtained in the capitalist economy are also insufficient for survival. Gregory and Piché state:

> Few are the households which can survive on their subsistence production alone, few are those which can survive on their income from the capitalist sector. This is the very nature of peripheral capitalist underdevelopment: "neither on foot nor on horseback." (1981, p. 21)

Capital benefits from this assymetrical articulation because the costs of the reproduction of labor power are absorbed in large part in the non-capitalist domestic mode.[7]

From this understanding of the nature of peripheral capitalist economies, Gregory and Piché make explicit the implication for biological reproduction. They assert that the household develops a collective strategy whereby some household members engage in wage labor (migrant labor), while other members participate in subsistence production and petty commercial production. Large family size (high fertility) is seen as an economically rational and essential response to this need for a labor supply for both domestic (household) and capitalist production. This is what Gregory and Piché refer to as a "double pressure for fertility" (1981, p. 27). Given that high fertility is seen as economic strategy, and in fact, necessary for survival, they argue accordingly that lower fertility would most likely result in greater impoverishment (1981, p. 28).[8]

Caldwell, in contrast to Gregory and Piché, provides an explanation of fertility behaviour and patterns which proceeds from modernization theory. For Caldwell, familial relations and their transformation are theoretically central, and he asserts that social Westernization is a more important precipitator of fertility decline than changes in the material conditions of people's lives.

According to Caldwell, the crucial determinant of high or low fertility is the magnitude and direction of "intergenerational wealth flows." Wealth is defined broadly, inclusive of goods, labor, services, protection, guarantees, and social and political support (1978, p. 573). Caldwell designates two fertility regimes which correspond with the direction of these wealth flows: a regime of unrestricted or high fertility, where the wealth flows from the younger to older generation (traditional society), and a regime of controlled or low fertility, where the wealth flows from parents to children (transitional society).

Traditional or pretransitional societies are characterized by what Caldwell refers to as a familial mode of production (comparable to Gregory and Piché's domestic or household production in its *precapitalist* form). The extended family is predominant, the conjugal bond is weak and authority is invested in elder males. High fertility persists in these traditional societies, according to Caldwell, because of the material advantages which accrue to the elderly males by virtue of their situational advantage. Specifically, it is the patriarchal control of consumption and familial labor that explains the persistence of high fertility. The patriarch's economic power and control extends to reproductive control, a control exercised in favour of high fertility because patriarchs reap the material benefits of a large number of children.

Consistent with his explanation of high fertility in pretransitional societies, Caldwell argues that fertility begins to decline when the flow of wealth is reversed, i.e., when parents spend increasingly on their children and both demand and receive very little in return (1976, p. 587).[9] The reversal of the flow of wealth is explained in terms of the emotional and economic nucleation of the family—the importation of the Western concept of family relationships and obligations. This entails, according to Caldwell, greater age and sex equality, loosened ties with extended kin and a closer and more sentimental conjugal bond (1978, p. 577).

Finally, Caldwell suggests that the adoption of the Western nuclear family form with a subsequent decline in fertility will occur irrespective of any economic changes: "fertility decline in the Third World is not dependent on the spread of industrialization or even on the rate of economic development. . . . Fertility decline is more likely to precede industrialization and to help bring it about than to follow it" (1976, p. 358). He asserts that the Western family is promoted by both "example and viewpoint" in both the mass media and in Western influenced education, and the attack on traditional familial relations is so strong that slower population growth is guaranteed (1976, p. 360–64).

Critique

The conceptual problems which impose limitations on these explanations of fertility behaviour can be drawn out through a discussion of the gender division of labor, the 1981 costs and benefits of children and patriarchy. In the Gregory and Piché model, patriarchy, or the domination of men over women, does not seem to exist in contemporary African social formations.[10] In the Caldwell model, on the other hand, women are portrayed as passive victims of patriarchal control, a control which is presumably relinquished with the adoption of the Western nuclear family form. In fact, neither formulation of the social relations of reproduction is adequate. It is problematic to assume that men have complete control over women's fertility and that biological reproductive decisions are always made for men's exclusive benefit. It is also problematic to assume that household members share equally in reproductive decisions and that the costs and benefits of reproductive activities are equitably distributed.

Gregory and Piché (1981) explain fertility behaviour exclusively in terms of a collective household strategy, similar to the microeconomists' conceptualization of the household as a homogeneous unit seeking to maximise its utility. Although microeconomists assume a "natural" sexual division of labor in their model, men's and women's activities are differentiated and consideration is given to the consequences of these differences for fertility behaviour. For Gregory and Piché, on the other hand, the gender division of labor is seemingly inconsequential. They implicitly assume that all members, whether through subsistence production, petty cash crop production or migrant labor, contribute on an equal basis and that these contributions are equitably distributed. They further imply that all members of the household are affected equally by peripheral capitalism. No conflicts, no contradictions and no inequalities exist within the domestic household, and high fertility merely represents an element of this collective household strategy.

The failure to incorporate the gender division of labor (and this includes human reproduction!) renders theoretically irrelevant the differential costs to members of their various activities and contributions. Further, the exclusion of women's subordination obscures the fact that women shoulder a disproportionate share of the costs of both childbearing and childrearing. In most African societies women carry the burden of childcare in terms of both time and cost, and with limited assistance from husbands (Adepoju 1977; Deere 1976; Huntington 1975; Due and Summary 1983; Monsted 1977; Roberts 1984; Stamp 1975). Even in extended families, where children are most often cared for by others of the group, the biological mother tends to care for infants under the age of three. As a consequence, high fertility increases the biological mother's responsibilities (Ferguson and Folbre 1981; Folbre 1983). As well, in terms of actual childbearing, frequent pregnancies in the face of heavy workloads, poverty and inadequate medical attention take their physical toll.

Rather than draw attention to the costs of childbearing and childrearing and attempt to theoretically interpret the demographic consequences, Gregory and Piché emphasize the productivity of children. The issue of costs is not simply a feminist insertion of little significance to our understanding of the dynamics of fertility patterns and behaviour. Given that these costs are absorbed primarily by women, the demographic implication is that men (husbands, partners) *may* be more likely to favor large families. In fact, some evidence suggests that wives in Africa do desire fewer children than their husbands (Belcher et al. 1978; Mack 1978; Okore 1977; Ukaegbu 1977). Albeit limited, this evidence indicates the need to analyse men and women's preferences separately and to theorize the gender dynamics related to human reproduction. In the policy-oriented and micro-level population literature, separate data collection for male and female members has already been stressed (Anker, Buvenic and Youssef, eds., 1982; Birdsall 1976; Tangri 1976). Theoretically, a conceptualization of the household as a homogeneous unit, as in Gregory and Piché's model, does not allow gender differences and unities to be explored.

It has been suggested that husbands may desire more children than their wives because the costs of reproductive work are not equally shared—or conversely, that wives may desire fewer children than husbands because they bear the brunt of the costs. It is highly probable, however, that the intrahousehold dynamics are considerably more complex. Gender differences in desired family size may also suggest that the material base to fertility may differ for male and female household members or that something other than a material base is in question. That is, high fertility may represent something more than a response to the demand for labor (perhaps some ideological dimension that addresses male virility?). Because gender differences have been ignored in empirical demographic work, the initial theoretical models and analyses which incorporate such differences are necessarily exploratory and speculative. Nevertheless, a prerequisite for any demographic theory is a recognition that gender differences may exist. This recognition, alongside the inclusion of the gender division of labor (which serves to direct attention to the costs of reproduction), should enhance a more complete understanding of the social relations of reproduction.

It is equally important not to neglect the productive value of children and the interrelatedness of their costs and benefits. While not immediately of economic value to the household, children eventually provide crucial labor contributions in many developing countries, including African countries. In this vein, Gregory and Piché's model of the "double pressure for fertility," a pressure which stems from the need to supply labor to participate in both domestic and capitalist production, is not only insightful but is also a real theoretical advance. As with the issue of costs, however, the benefits and value of children need to be further theorized in terms of possible gender differences and in light of women's subordination. Unfortunately, Gregory and Piché assume that a large number of children benefit male and female household members in the same way and to the same degree.

Some forms of assistance from children, such as economic support in old age, usually benefit both husband and wife. In other forms, there may be important gender differences. In areas where migrant labor is prevalent, women may directly benefit from high fertility in that children provide needed labor in subsistence production. And given that women are responsible for a combination of productive and reproductive activities, children can assist women with childcare to partially alleviate heavy workloads. In this context, differences between husband and wife may emerge over preferences for male and female children. The demographic literature commonly proclaims the preference for sons. While it is important for women to bear at least one son in African societies, ignoring daughters' labor is subtly sexist. Given the stringent gender division of labor in African societies, both sons and daughters make valuable contributions, and it may be that women desire more female children than men because daughters assist them in reproductive work.

The issue of the gender distribution of the *fruits* of women and children's labor must also be faced however. Husbands may favour a larger number of children than their wives not only because they share fewer of the costs, but also because they reap the greatest proportion of the benefits. Such benefits can encompass more than, for example, the appropriation of income earned from petty cash crop production. It can also include, as Hartmann (1981) has argued, personalized services and a higher standard of living with luxury consumption and leisure time. In a specifically demographic context, Caldwell postulates that men benefit from high fertility through increased leisure and consumption resulting from children's labor.

For Gregory and Piché what is central is capital's exploitation and domination of the domestic household and its members. One can infer from their argument that the exploitation from without (by capital) leaves no room for inequality and exploitation from within the domestic household. Capital, however, is not the only exploiter. The analysis or demographic model should expand to include exploitation within the domestic household. This can be partially achieved through an examination of the differential distribution of the benefits accruing from a large number of children. But consideration of decision-making power in the household and the resolution of gender differences, if and when they arise, are also needed. This perhaps brings us more directly to the issue of patriarchy. In order to adequately discuss Gregory and Piché's treatment of patriarchy, it is necessary to briefly review their historical examination of the domestic or household mode of production.

In their outline of precolonial African societies, Gregory and Piché follow closely Claude Meillassoux's theoretical model. The precapitalist domestic mode is characterized by male elders' control over the labor of children and control over the exchange of women and their fertility. The exploitative relations between the generations and the sexes are posited as necessary in order for the domestic community to reproduce itself.[11] Women play a central role by contributing to the daily reproduction or maintenance of the household's labor power and through biological reproduction.

Meillassoux does not hint at conflict or contradiction within this posited patriarchal society. Women's subordination is simply not problematic. Edholm, Harris and Young (1977), in their criticism of unhistorical, atemporal analyses of women, state:

> The problem seems exemplified by Engels, whose proclamation of a "world historical defeat of the female sex" virtually ends his discussion of women, and by Meillassoux, who seems to suggest that women were at some (very questionable) evolutionary stage simply raped and battered into subjugation, to remain forever mute and unprotesting (p. 119).

To their credit, Gregory and Piché assert that the control of the ruling patriarchy was not necessarily a smooth, nonproblematic process:

> The existence, for example, of a ruling patriarchy that appropriated power, and sometimes surplus production, for its personal benefit was the focus of considerable conflict between the generations and sexes (1982, p. 196).

Curiously, however, the conflicts and contradictions alluded to in the precapitalist domestic economy, and indeed the ruling patriarchy itself, are nonexistent in Gregory and Piché's conceptualization of the domestic economy in peripheral capitalist African social formations.

There is, however, no theoretical or empirical rationale to suggest that patriarchy disappears with peripheral capitalist development, and as Rhoda

Howard (1984) states, the macrosocial global inequalities do not render gender inequalities irrelevant (p. 46). By the same token, age and gender inequalities need to be addressed in conjunction with these global inequalities, or more specifically, in terms of the nature of peripheral capitalism. This is one of the ways in which Caldwell's work is problematic. While arguably providing a more perceptive treatment of the social relations of human reproduction, his analysis is also guided by demographic transition theory and the closely aligned modernization school of development.

In contrast to Gregory and Piché, the exploitative and hierarchical relations between generations and sexes are central to Caldwell's explanation of high fertility. The material benefits which accrue to the ruling patriarchy are the very basis for large family sizes. As in Meillassoux' formulation of the precapitalist domestic economy, the subordination of women and the patriarch's economic power and control over women's biological reproductive capacities are smooth and without conflict and contradictions. Women are passive, always acted upon rather than agents attempting to act in their own interests. While Gregory and Piché assume that a large number of children benefit men and women equally, Caldwell assumes that many children only benefit the ruling patriarch. This is clearly a matter for further investigation. Nonetheless, the suggestion that only the ruling patriarch benefits from a large number of children seems improbable. As proposed earlier, some benefits such as old age security accrue to both husband and wife, and women themselves may have their own incentives for a large family. In addition to the labor provided by sons and daughters there is the ideological dimension. The values of prestige, virility and fecundity usually surround human reproduction, and women as well as men may desire the prestige associated with many children.

Gregory and Piché, because of their conceptualization of childbearing as the reproduction of labor power, fail to consider the ideological dimension. Caldwell, on the other hand, defines wealth and benefits more broadly to include status and prestige. The problem is that he seems to assume that the production of a large number of children only enhances the patriarch's status. In African societies however, childbearing remains an important (and often only) source of status and power for women. By neglecting how women also benefit from a large family Caldwell overstates his argument and in the process excludes some further dynamics.

Like Meillassoux, Caldwell also presents women as acted upon, as victims subjected to complete domination and control by a ruling patriarchy. As with Gregory and Piché's model of contemporary African social formations, this analysis ignores the possibility of conflicts and contradictions surrounding human reproduction and gender relations. While men wield control over women's reproductive activities and capacities in a number of ways, nevertheless fertility may be one area where *complete* domination and control are more difficult to exercise. Women may resist and assert themselves in various ways: through illegal abortions, through the use of contraception without the husbands' knowledge; or possibly through some cultural equiv- alent to the "I have a headache" adage.

Further weaknesses in Caldwell's explanation of fertility patterns and behaviour emerge in his analysis of the transformation of familial relations and a postulated subsequent decline in fertility. It is questionable whether social Westernization, no matter how powerful or intense, is sufficient to transform either familial relations or levels of fertility in underdeveloped African countries. In making this argument, Caldwell reveals certain biases and displays an inadequate understanding of the nature of peripheral capitalist development.

It is not my intention to thoroughly critique demographic transition theory. For the purposes of this paper a few comments should suffice. The adoption of the Western nuclear family form is only available as an option to couples if they have the economic resources both to loosen ties with extended kin and to spend increasing amounts of money on their children, expecting little in return (Cain 1982; Thadani 1978). Further, Caldwell's "demographic innovators" are the urban elite who have been fully integrated into the capitalist mode of production. Most people however, as Gregory and Piché have pointed out, have to participate in both domestic and capitalist production. The very nature of peripheral capitalism demands the continued coexistence of the two modes, with the capitalist mode as dominant. It is highly unlikely, therefore, that familial relations will be transformed to any substantial degree in contemporary African social formations and concomitantly, unlikely that fertility will decline.

Finally, Caldwell's vision of the nuclear family and gender relations within it seems to be guided more by idealism than by a realistic appraisal. Patriarchal control over women's fertility presumably disappears because the conjugal bond becomes closer and more sentimental. This sentimental conjugal bond, asserts Caldwell, leads to improved negotiations between husband and wife about sexual relations and contraception which by some unknown logic decreases men's material advantages: "The very discussion of contraception and reproductive decisions almost certainly does something to lessen sex differentials and hence male material advantage" (1978, p. 576).

A more plausible interpretation is that men become more amenable to greater equality in fertility decision-making when or after the material and ideological benefits of children lessen. If most households became fully integrated into the capitalist mode, and if the nuclear family were constructed as Caldwell appears to view it, i.e., with a male wage earner and financially dependent wife and children, then there would be strong economic incentives for men to prefer smaller families. In this regard, Nancy Folbre (1983) has argued that the growth of wage labor weakens patriarchal control over adult children because independent income sources become available. This is not to say that the wealth flows from children to elderly males are immediately eroded, but their predictability declines. Once again, however, with the limited growth of full-time wage labor and the coexistence of the domestic and capitalist modes of production in contemporary African societies, there seems little likelihood these wealth flows will reverse.

In the nuclear family, Caldwell exaggerates the degree of equality between husband and wife. Although some women migrate to towns due to dissatisfaction and limited resources within the extended family (Obbo 1980), the nuclear family often decreases African women's economic independence. Their economic dependence in turn makes them more vulnerable to male demands and control (Van Allen 1974; Smock 1977). Patriarchy or the domination of men over women does not disappear in the nuclear family as Caldwell suggests but instead merely changes form.

From a feminist perspective the strength of Caldwell's work rests with his acknowledgement of women's inequality and his attempt to theorize the demographic consequences of patriarchy. Despite the problems with his particular formulation, Caldwell draws attention to some important issues surrounding the dynamics of the social relations of human reproduction which are most often ignored by demographers.

Conclusion

In this paper, I have attempted to clarify both the relevance of demographic theories for the further development of feminist theory and the relevance of feminist insights and scholarship for theories of demography. In the main, I have attempted to outline the problems which arise when a theory of demography is not informed by feminism. In a number of ways I have provided no definitive answers. I cannot offer a grand demographic theory which neatly incorporates the gender division of labor, gender differences in relation to desired family size and differential costs and benefits and women's subordination. What I hope I have shown is the need for demographic theories to address the issue of patriarchy and to consider the conflicts and contradictions of the social processes of human reproduction. In broad contours, what is needed is a demographic theory of fertility in developing countries which not only takes into account the global inequalities and nature of peripheral capitalist social formations, but also considers the inner dynamics of the household and hence gender inequalities.

Notes

1. The three basic and commonly accepted elements of population or demographic processes are mortality, migration and fertility. Admittedly, no discussion is complete

without the inclusion of all three. Nevertheless, this paper is confined to explanations of fertility.

2. Marxists generally view the Malthusian perspective as nothing more than apologetics for problems inherent in capitalist development—problems such as unemployment, rapid rural-urban migration, poverty, etc. For critiques of Malthusian analyses and the ideological nature of this perspective, see: Bondestam and Bergstrom, eds. (1980); Giminez (1977); Green (1982); Harvey (1981); Park (1974). See Seccombe (1983) for a critical appraisal of Marxists' rather sterile response to Malthusianism.

3. As an alternative to the modernization school of thought Seccombe (1983) has provided a reinterpretation of the demographic transition of mid-eighteenth to early twentieth centuries in Western Europe. In his introduction he eloquently argues for the need to integrate feminist and Marxist analyses into the field of demography and introduces some important revisions and correctives. In his own model, however, Seccombe surprisingly examines and explains demographic changes in terms of family forms and the reproductive *couple*. In this way, he has adopted one of the most problematic assumptions of conventional demographic theory—an assumption which theoretically excludes gender differences, conflicts and inequalities.

4. Maternal mortality statistics are not usually collected for women in Africa or in developing countries in general. See Harrington (1983) for a rare empirical study of Nigerian women and the detrimental health effects of childbearing and lactation.

5. In relation to developing countries, Mamdani (1972) was one of the first to utilize Marxist categories to explain population processes and behaviour. In his study he stresses the economic rationale for large families. Most of the criticisms which will be levelled at Gregory and Piché's model apply equally to Mamdani's study of Manpur in India.

6. Gregory and Piché do not, however, see this school of Marxist thought as incompatible with dependency theory (1981, p. 12–13).

7. Deere (1976) was one of the first to stress the fact that it is women who bear most of the costs of reproduction in peripheral capitalist social formations.

8. This essay was written before the appearance of Dennis Cordell and Joel Gregory, eds., *African Population and Capitalism: Historical Perspectives*, African Modernisation and Development Series, Westview Press, (1987). In this most recent work, Gregory, Cordell and Piché have moved some distance toward the view that household reproductive strategies are not simply collective, but are determined by a "patriarchal gerontocracy." The strategies are, further "embedded in intrahousehold power relations," and "are the result of conflictual as well as cooperative relationships." (Introduction, p. 31). But these general statements are not analysed in any depth, and despite them the basic theoretical dilemma remains: reproduction tends to be seen as subordinate to production, the power of capital as more important than the power of men over women, and class as more important than sex. Production of human beings assumes significance only because it is part of the constitution of the labor force.

9. Caldwell concentrates on the intergenerational wealth flows and attaches no theoretical importance to wealth flows from spouse to spouse.

10. In the 1987 formulation, as noted, it exists but does not seem to have much causal importance.

11. In the 1987 statement, Cordell, Gregory and Piché still adhere to this position (p. 25). For an excellent critique of Meillassoux, see Mackintosh (1977).

References

Anker, R., M. Buvinic and N. Youseff, eds. (1982) *Women's Roles and Population Trends in the Third World*. London: Croom Helm.

Barrett, M. (1980) *Women's Oppression Today*. London: Verso Editions.

Bay, E.G., ed. (1982) *Women and Work in Africa*. Boulder: Westview Press.

Belcher, G., A. Neumann, S. Ofosu-Amaah and N.D. Blumenfeld (1978) "Attitudes Towards Family Size and Family Planning in Rural Ghana." *Journal of Biosocial Science* 10: 59–79.

Beneria, L. (1979) "Reproduction, Production and the Sexual Division of Labour," *Cambridge Journal of Economics* 3: 203–225.

Beneria, L. and G. Sen (1982) "Class and Gender Inequalities and Women's Role in Economic Development—Theoretical and Practical Implications," *Feminist Studies* 1: 157–175.

Bondestam, L. and S. Bergstrom, eds. (1980) *Poverty and Population Control*. London: Academic Press.

Birdsall, N. (1976) "Women and Population Studies: Review Essay," *Signs* 1, 3 (Spring): 699–712.

Boserup, E. (1970) *Woman's Role in Economic Development*. London: George Allen and Unwin.

Cain, M. (1982) "Perspectives on Family and Fertility in Developing Countries," *Population Studies* 36, 2: 159–176.

Caldwell, J. (1976) "Toward a Restatement of Demographic Transition Theory," *Population and Development Review* 2: 321–366.

_____ . ed. (1977) *The Persistence of High Fertility: Population Prospects in the Third World*. Canberra: Australian National University.

_____ . (1978) "A Theory of Fertility: From High Plateau to Destablization," *Population and Development Review* 4, 4: 553–577.

_____ . (1982) *Theory of Fertility Decline*. London: Academic Press.

Caldwell, J. and P. Caldwell, (1978) "The Achieved Small Family: Early Fertility Transition in an African City," *Studies in Family Planning* 9,11: 2–18.

Cordell, D. and J.W. Gregory, eds. (1987) *African Population and Capitalism: Historical Perspectives*. Boulder, Colorado: Westview.

Deere, C. (1976) "Rural Women's Subsistence Production in the Capitalist Periphery," *Review of Radical Political Economics* 8,1: 9–17.

Due, J. and R. Summary, (1979) "Constraints to Women and Development in Africa," *Illinois Agricultural Economics Staff Paper No. 79*. Illinois: Department of Agricultural Economics.

Edholm, F., O. Harris, and K. Young (1977)"Conceptualizing Women," *Critique of Anthropology* 9/10: 116–130.

Feldman, R. (1984) "Women's Groups and Women's Subordination: An Analysis of Policies Towards Rural Women in Kenya," *Review of African Political Economy* 27/28 (February): 67–85.

Folbre, N. (1983) "Of Patriarchy Born: The Political Economy of Fertility Decisions," *Feminist Studies* 9,2: 262-284.

Folbre, N. and Ferguson (1981) "The Unhappy Marriage of Patriarchy and Capitalism," in L. Sargent, ed., *Women and Revolution*. Montreal: Black Rose.

Franke, R. (1981) "Mode of Production and Population Patterns: Policy Implications for West African Development," *International Journal of Health Services* 11,3: 361-387.

Giminez, M. (1977) "Theories of Reproductive Behaviour: A Marxist Critique," *Review of Radical Political Economics*, 9: 17-25.

Green, E. (1982) "U.S. Population Policies, Development and the Rural Poor of Africa," *Journal of Modern African Studies* 20,1: 45-67.

Gregory, J.W. and V. Piché (1981) "The Demographic Process of Peripheral Capitalism Illustrated with African Examples," Montreal: McGill University Center for Developing Area Studies.

Gregory, J.W. and V. Piché (1982) "African Population: Reproduction for Whom," *Daedalus* 111,2 (Spring): 179-210.

Hafkin, N. and E. Bay, eds. (1976) *Women in Africa: Studies in Social and Economic Change*. Stanford: Stanford University Press.

Harrington, J. (1983) "Nutritional Stress and Economic Responsibility: A Study of Nigerian Women," in Buvinic et al., eds., *Women and Poverty in the Third World*. Baltimore: John Hopkins Press.

Hartmann, H. (1981) "The Unhappy Marriage Between Marxism and Feminism," in L. Sargent, ed., *Women and Revolution*. Montreal: Black Rose Press.

Harvey, D. (1974) "Ideology and Population Theory," *International Journal of Health Services* 4,4: 515-537.

Howard, R. (1984) "Women's Rights in English-speaking Sub-Saharan Africa," in C. Welch Jr. and R. Meltzer, eds., *Human Rights and Development in Africa*. Albany: State University of New York Press.

Huntington, S. (1978) "Issues in Woman's role in Economic Development: Critique and Alternatives," *Journal of Marriage and the Family* 40,4 (November): 1001-1012.

Mack, D. (1978) "Husbands and Wives in Lagos: The Effects of Socioeconomic Status on the Pattern of Family Living," *Journal of Marriage and the Family* 40,4 (November): 809-820.

McDonough, R. and R. Harrison (1978) "Patriarchy and Relations of Production," in A. Kuhn and A. Wolpe, eds., *Feminism and Materialism*. London: Routledge and Kegan Paul.

Mackintosh, M. (1977) "Reproduction and Patriarchy: A Critique of Claude Meillassoux, 'Femmes, Greniers et Capitaux,'" *Capital and Class* 3: 119-127.

Mamdani, M. (1972) *The Myth of Population Control: Family, Caste and Class in an Indian Village*. New York: Monthly Review Press.

Matsepe, I. (1977) "Underdevelopment and African Women," *Journal of Southern African Affairs* 11 (April): 135-144.

Meillassoux, C. (1981) *Maidens, Meal and Money: Capitalism and the Domestic Community*. Cambridge: Cambridge University Press.

Monsted, M. (1977) "The Changing Division of Labour Within Rural Families in Kenya," in Caldwell, ed. *The Persistence of High Fertility.* Canberra: Australian National University.

Mott, F. and S. Mott (1985) "Household Fertility Decisions in West Africa: A Comparison of Male and Female Survey Results," *Studies in Family Planning* 16,2: 88–97.

Obbo, C. (1980) *African Women: Their Struggle for Economic Independence.* London: Zed Press.

Okore, A.O. (1977) "The Ibos of Arochukwu in Imo State, Nigeria" in Caldwell, ed., *The Persistence of High Fertility* .

Park, R. (1974) "Not Better Lives, Just Fewer People: The Ideology of Population Control," *International Journal of Health Services* 4(4): 691–700.

Remy, D. (1975) "Underdevelopment and the Experience of Women: A Nigerian Case Study," in R. Reiter, ed., *Toward an Anthropology of Women.* New York: Monthly Review Press.

Roberts, P. (1984) "Feminism in Africa: Feminism and Africa," *Review of African Political Economy* 27/28: 175–184.

Rowbotham, S. (1973) *Women's Consciousness, Men's World.* Harmondsworth: Penguin.

Sargent, L., ed. (1981) *Women and Revolution.* Montreal: Black Rose.

Seccombe, W. (1983) "Marxism and Demography," *New Left Review* 137: 22–47.

Smock, A.C. (1977) "The Impact of Modernization on Women's Position in the Family in Ghana," in A. Schlegel, ed., *Sexual Stratification: A Cross-Cultural View.* New York: Columbia Univeristy Press.

Stamp, P. (1975–76) "Perceptions of Change and Economic Strategy Among Kikuyu Women of Miterok Kenya," *Rural Africana* 29 (Winter): 19–44.

Staudt, K. (1975–76) "Women Farmers and Inequities in Agricultural Services," *Rural Africana* 29 (Winter): 81–94.

Strobel, M. (1979) *Muslim Women in Mombasa, 1890–1975.* New Haven, Conn.: Yale Press.

Tangri, S. (1976) "A Feminist Perspective on Some Ethical Issues in Population Programmes," *Signs* 1, 4.

Thadani, V. (1978) "The Logic Sentiment; The Family and Social Change," *Population and Development Review* 4, 3: 457–499.

Tinker, L. and M. Bramsen, eds. (1976) *Women and World Development.* Washington: Overseas Development Council,

Ukaegbu, A.O. (1977) "Family Planning Attitudes and Practices in Rural Eastern Nigeria," *Studies in Family Planning* 177–183.

UN Economic Commission for Africa. (1972) "Women: The Neglected Human Resource for African Development," *Canadian Journal of African Studies* 6,2: 359–370.

UN Economic Commission for Africa (1975), "Women and National Development in African Countries: Some Profound Contradictions," *African Studies Review* 18,3: 45–70.

Van Allen, J. (1974) "Women in Africa: Modernization Means More Dependency," *The Centre Magazine,* (May/June): 60–67.

5

Rural Women's Access to Labor in West Africa

Penelope A. Roberts

Introduction

Richards concluded from a review of farming systems research in Africa that:

> Perhaps the most important aspect of recent reassessments of African peasant agriculture is the realisation that shortage of labour is often a greater constraint on production than shortage of land (Richards 1983, p. 30).

Access to labor, choice of agricultural practices in the light of constraints of labor availability and minimization of fluctuations in labor supply have been and are organizational issues of the greatest importance to farmers. Much of the rural labor in West Africa is recruited through the non-market relations of household membership, kinship or socio-political rank. Even contracts for wage labor may be modified by these relations. Access to labor, therefore, is largely dependent upon the hierarchies of gender, rank, generation and class within and between households. Women are, on the whole, profoundly disadvantaged within these hierarchies. This has significant consequences for the way in which women enter into production.

In her comparative study of "female" farming systems in Cameroun and of "male" farming systems in Western Nigeria, Guyer (1984a) demonstrated the ways in which agricultural practices have been modified as a consequence of the differential access of women and men to labor. Women farmers with

---

I should like to thank the many colleagues who have made helpful comments on earlier versions of this paper, but particularly Ann Whitehead and Gavin Williams. This paper is intended to be tentative and exploratory. Rather than including notes on every doubt, hesitation and qualification which would exceed the length of the paper, I have included none.

restricted and unreliable access to the labor of others operated their farms to avoid, for example, labor bottlenecks at peak periods. Male farmers with greater resources for the mobilization of labor retained periods of labor with greater resources for the mobilization of labor retained periods of labor intensity during the farming year. The long hours that women farmers work on the fields, in crop processing and on domestic chores may well be attributed more generally to women's lack of access to the labor of others; to their inability to share with or delegate work to kin or other members of the household or to recruit labor from the rural labor market.

In this paper, I want to examine and extend the arguments of Guyer and others concerning the constraints on women's access to labor. The issue concerns the social relations of production, particularly those of gender, which form the basis of the differential access of women and men to the labor of others. These relations determine the ways in which women and men can recruit and mobilize labor. The questions may be stated as follows. First, in what circumstances do women recruit the labor of others rather than being the subjects of labor recruitment themselves? I shall approach this question by examining the relationship between "household" production and women's own-account enterprises. Secondly, which women can and do recruit the labor of others and whose labor do they mobilize?

These questions have a practical and political purpose. If it is the case that lack of access to labor rather than lack of access to land is more likely to be a constraint upon production, especially where women farmers are concerned, then various aspects of agrarian reform and of rural development are being misconceived. For example, guaranteeing women's independent rights in land so that widows or the divorced retain their means of existence is insufficient if they have no means of recruiting labor to work it. The fact that female headed households are so frequently classified amongst the poorest of the poor can often be attributed just to that fact. These circumstances have affected women's entry into and management of non-farming as well as farming enterprises although this paper concentrates on the latter. Rural women engage in a variety of own account enterprises such as trading, crop-processing and craft production. The constraints on the allocation of labor to these enterprises are the same as those which affect farming, and similar patterns of labor use can often be discerned. Very often, however, development projects for women which promote these activities, such as craft production, utterly fail to recognise this major constraint. Women's own account enterprises compete for labor from within and outside the household. On the one hand, there are claims upon the labor of the woman herself: she can only work "part-time" on her own-account enterprises. On the other, there are prior claims upon the labor she might wish to command.

The Differential Access of Women and Men to Labor

One might conclude from a review of the literature on women farmers in West Africa that they either work for others or alone, with no assistance, single-handedly. The former case includes systems of "household production" in which women work under the authority of male household heads/ husbands. The latter case includes, on the one hand, the "female farming system" in which women produce most of the crops designated for household consumption and, on the other, "own account" enterprises in which women produce or acquire access to farm or livestock products which they may dispose of independently of the household. Women can be involved in both forms of production. They may provide labor in the household farm and may also, for example, have their own private plots which they work on their own. It is rare to find reference to systems in which women are perceived to initiate the process of production through the mobilization of labor from household, kin or the socio-political networks of the community. It is even rare to find reference to women employing wage-labor although it is at least usually explained why, despite their labor needs, they do not; they cannot afford to do so. With the most common exception of the help of children and sometimes aged mothers, most enterprises seem to be operated single-handedly.

Much of this literature takes for granted what I hope to explain: why do women expect to make farms and manage "own-account" enterprises with so few occasions for mobilizing the labor of others and, therefore, with such limited opportunities for accumulation? I argue that these are not the consequences of any natural tendency on the part of women to depend, at best, only on the labor of children and the aged: to be drudges on their own account as well as on that of their husbands! The labor processes involved in the management of women's enterprises are products of history, of the transformation of gender and class relations. In so far, however, as this natural tendency is taken for granted, it is possible that women have mobilized labor or do now do so in ways which remain invisible. An examination of the historical circumstances in which women have been able to mobilize labor clarifies the contemporary constraints on the vast majority who do not.

There are at least two schools of thought which have contributed to the dominant impression that women either work for others (men) or on their own. First, there has been the influence of neo-Marxism, particularly in respect of debates around production and reproduction prior to the relocation of these debates around gender (see, for example, Meillassoux 1975). I shall not dwell on these except to mention that the concept of "control over women," as producers (labor) and as reproducers (of future labor) effectively denied consideration of women's own capacity to mobilize the labor of others. Secondly, there has been the influence of what Guyer (1984a) has described—and decisively criticised—as "naturalism" in models of African production.

The "naturalism" model, originating in the work of Baumann (1928) and subsequently elaborated by Boserup (1970), maintains that "female farming systems" are "primitive" in the evolutionary sense. They are associated with root crop cultivation and characterized by a low intensity of labor input and the absence of peak periods of labor use. A single person is able to carry out all, or most, of the tasks in sequence. Farmers did not, therefore, need to mobilize the labor of others. The introduction of cereals requiring more intensive methods of production led to the increasing involvement of male labor and progression towards "male farming systems."

The term "male farming systems" conceals, of course, the use of women's labor in such systems of production (Beneria 1981). But the main problem with the alleged difference between female and male farming systems is that the model is based on the assumption that the crop itself determines the labor practices associated with its cultivation. Guyer demonstrates that the "individuated" systems of production are not associated solely with root crops and therefore with female farming systems. Rather, they are associated with the new staple crops (maize, cassava). The cultivation of the old staples (yam, millet, sorghum) is more frequently associated with elaborate, ritualized, divisions of labor, an "interdigitation of tasks" which requires the sequential and interdependent use of the labor of women and men, young and old, and a continual legitimation of rights in the labor of those involved (Guyer 1984a, p. 383). Critically, it is men of seniority in household, lineage or community who may initiate these social forms of production. She argues, therefore, that "female farming systems" did not precede "male farming systems" in any evolutionary sense nor are the different systems of production determined by the crop. The individuated systems of cultivation associated with female farming are the consequence of a major and fairly recent transformation in the organization of production.

The naturalist model also claims an association between the low intensity of labor in women's farming systems and the constraints of child-bearing and rearing. This requires women to "spread" their labor input rather than incorporate periods of greater labor intensity during the farming year. Guyer shows, however, that male farmers retain periods of labor intensity *because they mobilize the labor of others, while women farm largely on their own despite their responsibilities for domestic labor.* Guyer, therefore, insists that the main difference between "female farming systems" and "male farming systems" lies in the differential access to labor of women compared to men.

In her comparative study of the male farming systems of the Yoruba in Western Nigeria and of the female farming systems of the Beti in Cameroun, Guyer concludes that "there are qualitatively different kinds of power which activate production" and that power to activate production differs significantly between women and men (Guyer 1984a, p. 384; see also Guyer 1984b). Men draw on their authority and seniority in household, lineage and community and recruit male and female labor for their farming enterprises. Even their employment of wage labor draws on community standing in respect of credit and well established standards of work and rates of pay which reflect the social standing between employer and labor. As a result, there are considerable differences between men in their capacity to mobilize labor and this constitutes a potential basis for class formation. Beti women farmers, on the other hand, lack the authority to recruit either male or female labor in any of these institutions. The labor contribution of male members of their households is "erratic" and cannot be built into the management of labor practices on their farms. There are no established standards of work and rates of pay for wage labor employed by women farmers. There is "no accepted sanction for women in their attempt to recruit labour . . . the recruitment of labour by women does not draw on established institutions" (Guyer 1984a, p. 380). There are few differences between women in respect of these constraints and, consequently, little basis for socio-economic differentiation and class formation between them.

Guyer's studies of the male and female farming systems can be compared to a number of other studies which have described women's social incapacity to initiate production through the mobilization of the labor of others in their own account enterprises in "male farming systems." Hirschon (1984) argues that "property, in some societies, may best be thought of in terms of labour." Such societies include the kin-based societies of West Africa where rights in land can be obtained through a variety of social and political affiliations, including those of being a wife. However, rights in land cannot be isolated from the social and political hierarchies from which they are derived and "the usefulness of land rights is limited by the extent of rights in labour (and) are more or less meaningless without the labour power to work it" (Whitehead 1984, p. 184). Whitehead's study is of the Kusasi of north east Ghana where production, initiated by male household heads, mobilizes the labor of all subordinate members of the household and, depending on their political status in clan and community, of labor from outside the household. In addition, junior men and all wives have rights to farm private plots of land from which the product is solely theirs to dispose of. Women's plots are, however, very much smaller than those of men. This is partly because, as in marrying wives, they lack the social and political affiliations through which to activate land rights. The chief cause, however, is their lack of social power in the hierarchies of household, domestic group, clan or community which activates access to labor. Adult women were unable to command the labor of social superiors, either within or without the household:

[women] could not command the labour of adult men. In terms of gender, the net effect is that Kusasi women utilise both household and exchange labour differently from men. It is difficult for a woman to command the labour of social superiors, either within the immediate polygynous family unit or within the set of neighbouring agnatically related households. They rely on very junior men, often young boys. They never mount the large exchange work parties which are so important for male farmers, especially household heads (Whitehead 1984, p. 184).

A similar situation has been observed in the Hausa areas of southern Niger (Roberts 1977 and 1986). Wives farm their private plots (*gayamna*) virtually single-handedly in the time left free from labor on the household fields. They have, in consequence, adopted farming practices quite different from those used on the larger household fields. They plant more densely to avoid weeding and use manure when they can get it. There is, however, an intense struggle going on between husbands and wives, household heads and subordinate labor, over the allocation of labor between household fields and women's private plots. This struggle has its basis, in part, in the declining control of men over subordinate male household labor. Central to this issue is the relationship between household production and women's own-account enterprises.

Household Enterprises and
Women's Own Account Enterprises

The studies cited above locate the power to mobilize labor for production within the hierarchies of gender, generation, rank and class within and between households, in all of which women are subordinate. In the case of Hausa and Kusasi women farmers in "male farming systems," these conditions of household production affect women's capacity to enter into own account production and I examine the relationship between these first. Secondly, I consider the question of whether all women suffer to the same degree from inadequate access to labor. Guyer concluded that, in the Beti case, women were all more or less equally disadvantaged but this is not a conclusion which must be reached in all circumstances. Women are not necessarily equally subordinate within all these hierarchies. At least historically we can determine considerable differentiation between women specifically in respect of their capacity to mobilize labor.

The right of wives to operate own-account enterprises has been widely reported in West Africa. It has generally been accounted for as a correlate of the "separate purse" of husbands and wives in polygynous, lineage based societies. The "separate purse" allows spouses to meet their own obligations to kin groups. So well established has this correlation become that it has tended to lose sight of the material conditions required for women to enter into own-account production, which include access to land (in the case of farming) or money capital and credit (in the case of trading) and, above all, the opportunity for a woman to dispose of some part of her own labor time and perhaps that of others. There has, in fact, been a tendency to reify the institution: to designate it as a traditional practice which distinguishes conjugal relations in Africa from, for example, the "joint purse" or "conjugal fund" of European monogamous marriage (Goody 1976). It has even been represented as evidence for the relatively greater equality between women and men in precapitalist African societies than in those of Europe. Boserup, for example, supports her argument on this point on the presence of this institution. She makes, however, curiously limited claims for how it operated to women's advantage:

> The women are hard-working and have only a limited right of support from their husbands, but they often enjoy considerable freedom of movement and some economic independence from the sale of their own crops (Boserup 1970, p. 50).

Boserup also concludes that even these limited advantages were not enjoyed by all women, either all the time or at all. They were, she suggested, a condition of the institution of polygyny which allocated the "most tiresome jobs" to junior wives whose status was "inferior as befits the assistant or even servant to the first wife" (Boserup 1970, pp. 43, 45). Further, she traced the institution of polygyny to that of domestic slavery and of pawning, particularly of women. If these institutions have been critical to women's own-account enterprises, then historically there have been very unequal opportunities for women's entry into own-account enterprises and changes in sources of labor should have modified these conditions considerably. The first problem, however, is that the notion that wives had limited support from their husbands but some degree of economic independence does not explain the contemporary relationship between husband's enterprises and wives own-account enterprises.

I have used the term "own-account enterprise" in lieu of any other. In contemporary practice, it covers a wide variety of activities by which women earn some income for their own use. The critical feature of these activities is that they arise from women's "rights" to dispose on their own account of a small part of the resources owned or produced by the household into which they are married, including some part of their own labor time. Those most commonly described include: (1) production on "private" plots—those generally allocated by a household head to his male and female subordinates, the product of which is not directly destined for household consumption and not under the immediate distributive authority of the household head;

(2) the sale of products of the wild or of crops or of animal by/products as the disposal of or produced by the household (including, of course, by women's own labor) sometimes referred to as "surplus" to household consumption; (3) the right to specific payments in kind for certain types of labor on household or husband's farms which provide a source of raw materials at women's disposal. These practices provide women with products which may be traded in their raw form or additional labor may be invested in processing (for example, cooking, preserving or craft manufacture) to add value. Income may, under certain conditions, be reinvested to expand these self-produced resources, such as purchasing additional crops.

It is essential to realize, first, that these contemporary enterprises derive from the status of women as wives/co-wives/sisters/mothers/widows/daughters and so forth in male headed households. Second, status as wife/mother/sister may determine the particular type of resources at the disposal of a woman. For example, a mother may be able to dispose of some part of the labor time of a daughter in her own-account enterprises. It follows that women's opportunities for own-account enterprises may change over her life time. Third, these "rights" are usually explicitly recognized (they may be cited, for example, in divorce cases) but this does not mean they are inalienable, unchangeable or never abused. Thus women's own-account enterprises are by no means autonomous of male headed households, but rather a condition of women's relationship to them and within them. One of the key limitations arising from these relationships is that women's own-account enterprises are subordinate to the household enterprise and sometimes at odds with its interests.

One limitation is that the claims of a male head of household or husband over his wives' labor services are considerable and are not reciprocal. The extent of wives' obligations to provide labor to their husbands is a major constraint upon the development of their own account enterprises. It is often a more insoluble constraint than that of gaining access to other means of production such as land, small amounts of capital and so on. Conversely, the claims which wives can make on their husbands' labor are invariably less and less secure. The relative strength or weaknesses of these claims, however, has to be located in the context of the relationship between husbands' and wives' enterprises and other sets of obligations arising within marriage. "Household production" may conceal unequal contributions in the form of labor and unequal distribution of the product of labor, whether in the form of consumption or accumulation of material or social status. When there is evidence of such discrimination against women, as wives, we may well wish to describe "household production" as a male enterprise based on the exploitation of wives' labor and the continued reproduction of gender hierarchy in the relations of production. However, wives do have material interests and claims upon such household/male enterprises including and exceeding those of personal consumption, notably the interests of their children.

It is on these grounds that women's own account undertakings appear conceptually distinct from household/male enterprises. Wives appear to dispose of the product or income of their undertakings free from claims by husbands, whereas they have specific claims to personal consumption and investment in their children from the household/male enterprise. Women's own-account ventures appear to be "autonomous" of the household in the sense that household members have no institutionalized rights in their product. They have usually, therefore, limited or no specific obligations to contribute labor to them. Moreover, the labor of wives in their own-account venture is seen as "surplus" to the labor requirements of the household, done in wives' "spare time" and, in that sense as well, autonomous of the management and organization of the household as the enterprise of all its members. In this respect, women's own-account enterprises may appear to be, and may be, in competition with those of the household in so far as some part of her labor time is withdrawn from its service.

The negotiations between women and men for access to each other's labor and for access to the resources in labor which each might control are invariably complex. Negotiations are embedded in "contracts" of a wider nature, such as marriage. These are invariably subject to renegotiation and to redefinition, but they circumscribe women's access to labor within and outside the household (Roberts 1986). A senior wife's access to the labor of a junior co-wife may be modified by husbands' demands on the junior wife's labor. A woman's access to work parties may perhaps be negotiated only through her husband. Access to the labor of other women in the community may be restricted by their conjugal or familial obligations. Access to the labor market may be limited, above all, by a woman's lack of cash. But it may also be limited by her lack of social power to sustain long term contracts and, sometimes, by the "familialization" of wage labor—its incorporation into the social hierarchy of the household. Women who lack obligations to husbands, such as widows or divorcees, may also lack the means of mobilizing labor on their own behalf.

Any attempt to describe which women have the means of mobilizing labor, and which labor they can mobilize, is bounded by these complex hierarchies. A woman's age, her rank in relation to wives in a polygynous marriage, the rank of her husband in a multi-generational patriarchal household, her status as inherited wife or widow, the number and sex of her children and of her children's spouses are one set of parameters. These relationships entail labor obligations of women to women, men to women and women to men, and vary from society to society.

Let us consider, for example, labor obligations between women. In some societies, ranking between wives involves significant sources of command over a junior wife's labor; in others, wives' labor (except that which they can reserve for their own-account enterprises) is only at the disposition of husbands and cooperation on a voluntary, and generally very limited basis, is the best that a co-wife can hope for. Junior wives may eventually substitute for the labor obligations of a senior wife to her husband, as I have argued in the case of the Yoruba (Roberts 1986). But they are under no obligation to provide labor services to the own-account enterprises of a senior wife. The labor of daughters, on the other hand, may be largely at the disposition of their mothers, but this obligation may be considerably reduced at marriage when a daughter enters her husband's household.

As far as women outside the household are concerned, the ability to mobilize their labor appears to be very restricted. Two circumstances recur. First, there may be reciprocal agreements between women to provide each other with labor on a turn and turn about basis. For example, in the Gambia women may organize work parties for harvesting each other's rice farms (Dey 1980). These, however, are non-hierarchical forms of labor mobilization involving the distribution of labor over time between women on an equal basis. On the other hand, wealthier or more powerful husbands may use their facilities for mobilizing work parties on behalf of their wives on a principle of reciprocity but a practice of inequality. As in the Kusasi case cited earlier, the wives of poor men or men with few social resources are more likely to work for others in such work parties than have one working for themselves.

It would seem, therefore, that women have limited access to the labor of other women. Authority derived from seniority of rank and generation within the household (daughters, daughters-in-law, co-wives) is modified by husband's authority over the labor of the same person. Women's power to mobilize the labor of women outside the household on a non-reciprocal basis is highly restricted, and largely dependent upon her husband's good will to exercise his own capacity in this respect on her behalf. This strategy depends on the male hierarchies of social and political authority: women recruit labor and are recruited as labor in such arrangements through their relationships to men.

The use of the income from women's own-account enterprises, however, is not so divorced from the interests of the household. The use of the term "own-account" is not intended to imply that income is devoted to personal consumption in the sense of the derogatory term "pin-money." Product or income from these enterprises contributes to the costs of reproduction of the household/male enterprise. Its contemporary use is sometimes specified in conjugal relations, such as a wife's obligation to provide certain types of food for family (especially child) consumption. In the longer term, investment in children may ensure security in a woman's old age. In the case where she has no children, it may reduce her vulnerability.

The use of women's income from these enterprises varies according to the contingent circumstances. In matrilineal societies, it may be invested in long term securities for the matrilineage. In patrilineal societies, a distinctive strategy recurs: women's enterprises often seem intended to recuperate, via the sale of products, the cash incomes of men (as husbands, male heads of household) acquired through their power to dispose of the bulk of income derived from the deployment of their dependents' labor.

These factors suggest that not only do women have to manage their enterprises under different constraints from those of men, but they also pursue different objectives in the realization of their own-account ventures than those realized in their participation in production in household/male enterprises. In circumstances, for example, where women are not in charge of the regular supply of staples to the household (though their labor is involved in their production), they may specialize in the production of crops, or crop varieties, not destined to the eventual or sole consumption of the household but for the market. For example, women specialize in the production of groundnuts and varieties of millet for use as seed stock or processed "snacks" in Niger. This strategy circumvents to some degree the limitations set by the very small areas of land they can farm on their own. It is probable that women's own-account farming enterprises are often innovative in such ways, inserting themselves into the agrarian economies as a consequence of these constraints and objectives.

Case studies describe various aspects of the relationship between "household" enterprises and women's own-account enterprises which I have suggested. In the Yoruba speaking areas of Western Nigeria, women do not normally farm on their own account but are entitled to pursue their own-account businesses and specialize in trade, especially in farm products. Trade is sometimes initiated from their rights to payment in kind for certain types of farm labor performed for their husbands (Clarke 1977). Husbands are expected to provide basic consumption needs to co-resident wife (wives) and their children, but mothers are expected to provide additional foods and regular subsidies to their children's welfare (Galletti 1956; Roberts 1977). However, as Berry's study shows, wives' obligation to provide labor services to their husbands restricts the amount of labor they can devote to their own trades. Male cocoa farmers expect their wives to provide labor while a farm matures though generally prefer to use hired labor once it is yielding an income. However, a wife can "still be called upon at any time to help with her husband's work" (Berry 1985, p. 95). A wife must assist in her husbands' trade or manage his productive cocoa farms while he establishes new ones elsewhere. Women do not, however, have access to their husband's labor nor can they even expect their husbands to contribute to the establishment of their own trade. The key issue is the contradiction between a husband's need for his wife's labor and the labor time a wife needs to devote to her trade in order to establish an income which will contribute sufficiently to her children's support and household expenses to compensate for the cost of replacement of her labor. Thus "a farmer's attitude towards his wife's economic independence was likely to be somewhat ambivalent" (Berry 1985, p. 124). As should be expected, since women can rarely deploy the labor of others to carry out any part of the labor involved in establishing their trade (though women sometimes help each other out), they work very long hours, normally establish their enterprises very slowly and often do not prosper.

In the case of Hausa Niger, which I have already mentioned, wives can draw on very little labor for assistance in production on their *gayamna* farms. They have adopted their farming practices and specialized in certain types of crop production to accommodate these constraints. The set of conflicting interests that arise are frequently articulated. Husbands are increasingly dependent upon wives' labor on household farms as their authority over subordinate male labor has declined through seasonal emigration. The pressure they place on wives to reduce the amount of time they spend in such production is mitigated, however, by one interest that men have—and may cynically express—in women's own-account farming enterprises. Men control the distribution of land to wives and in areas of land shortage may allocate poor quality land which women restore through careful use of manure and nitrogen restoring crops such as cowpeas. The land is then recuperated for male use. Conversely, the sale of crops women produce, especially those processed as snacks, recuperates into women's hands some part of the income that men acquired from their control of the disposal of crops from household fields produced through male access to female labor (Raynaut 1968; see also Guyer 1984b, p. 105).

A different situation arises in the matrilineal, cocoa producing areas of Ghana. Here women produce many of the basic food crops in a farming system which has been modified to accommodate the care of young cocoa plantations. A number of studies have addressed the relationship between household enterprises and women's own-account enterprises, in particular women's cocoa farms though women also specialize in a variety of other enterprises ranging from the collection and processing of forest products such as snails and loofahs to the production of soap and vegetable salt and trade in foodstuffs. Most of these studies show that wives' and childrens' obligations to work on their husbands'/fathers' cocoa farms leave women with little time to work on their own. Husbands have no institutionalised obligations to provide labor on their wives' farms. The limited income that women can obtain from their own account enterprises under these conditions means that they cannot either afford wage labor. Okali describes the agonizing choices between two limited strategies available to wives. One is to hope that eventually a husband will show his gratitude (not his obligation) to his wife either by paying for labor to work her farms or by giving her one of the farms they have made together. Nothing compels him to do either and indeed, since the attainment of a small income from her own farms is sufficient for many women to separate from or divorce their husbands, thus leading to the loss of her labor, the majority of husbands never do this. The alternative strategy is to divorce in any case and look for a more generous husband, or try to manage alone. As a consequence, wives have abandoned farms for lack of labor. The strategy of divorce or separation from husbands is a condition of the circumstance that a wife still has no customary right to inherit any part of her husband's cocoa farms. Although he must contribute to her maintenance during marriage, her contribution to her own subsistence and that of her—and his—children is considerable (Mikell 1984; Okali 1983; Oppong 1975; Roberts 1987; Vellenga 1977). Okali, indeed, shows that women are more likely to be own-account cocoa farmers if they are not married, or not living with their husbands (Okali 1983, p. 56).

Tradition or the Transformation of Gender and Class Relations?

I have suggested, above, various contemporary forms of the relationship between household enterprises and women's own-account enterprises. They are interdependent in the sense that a woman's resources may derive from her position in her husband's household, they compete over the allocation of the woman's own labor. The question is, what relationship do these have to women's supposed traditional right to operate own-account enterprises? In the present state of knowledge, only very tentative hypotheses can be suggested and I shall focus on only one here. This is that some women's enterprises in the past were able to operate on a larger scale, mobilizing the labor of others from outside the hierarchies of gender, rank and generation. This source of labor was that of slaves and pawns.

Given the hierarchal nature of gender relations within and outside the household which usually allocate more power to men in the mobilization of labor, a source of labor which was not immediately incorporated into those hierarchies would seem of critical importance to women. Robertson and Klein (1983) have made this point, arguing that "free women derived much of the benefit from slave labor in sub-Saharan Africa." In the case of male slaves this was because "in many societies women had greater difficulties than men did in recruiting male labor through the lineage system." Women slaves, who constituted the majority in West Africa, offered women owners the same advantages as they did to male owners: an increase in the quantity of labor at their disposal. But in addition, women owners benefitted from women slaves because "women did most of the agricultural and virtually all the domestic work. The value of women slaves was based on a sexual division of labor which assigned much of the productive labor to women" (Robertson and Klein 1983, pp. 11-13).

Access to labor of slaves and pawns provided women with the means to establish enterprises outside marriage and male headed households. These, then, were female headed households rather than own-account enterprises but ones in which women's inability to mobilize labor had been circumvented through the use of labor independent of gender hierarchies. Under some circumstances, this may have involved a woman's physical withdrawal from the structures of household and community, even if she was a woman of rank rather than subordinate within these structures as in the case of Ya Ndama of Sierra Leone (MacCormack 1983). Similarly, the remarkable success of Accra women's trading enterprises based on slave and pawn ownership may have been associated with the separate residences of women and men and the consequent fracture of gender hierarchies (Robertson 1983; Klein 1981).

Under other circumstances, this source of labor may have relieved women, as wives, of some of their labor obligations to husbands, and enabled them to pursue their enterprises from within the male-headed household, but not dependent upon it. An example of the possibilities available to free women deriving from the availability of slaves can be drawn from the Sefwi Wiawso area of Western Ghana in the late 19th century. Profits from the collection of wild rubber were invested in slaves which were bought as "means of reward for favorite wives and children, used as labour in gold digging, rubber production and snail collection" (Arhin 1972, p. 38). Women also bought slaves, male and female, for themselves (Roberts 1984). Women bought them especially for their snail collecting and processing enterprises which involved the management of considerable resources in labor and material.

Nevertheless, while slaves may have provided a valuable source of labor to free women, there were ambiguities about the nature of domestic slavery when wives remained co-resident in husbands' or male-headed households which qualified the advantages free women derived from their labor. These arise from the incorporation of many slaves into the gender hierarchy of the household. As Arhin says, slaves were "incorporated into families." The slow process of incorporation into the household (described in Perrot 1982, p. 165ff) gradually subordinated the slaves given to women or bought for them into the hierarchy of gender and generation in the household. No wonder, then, that the legend of women's freedom and autonomy in this matrilineal area is of a community without men. Such was Pomakrom, a village described now in legendary terms but which really did exist in the 1920s when Carinall visited it. Poma had founded the village in the late 19th century with one woman companion and had "made the rule that no man should marry either of them or any other girl who might join them; that men could stop only a while in the village, and would have no rights at all over any children born. . . . There were some two hundred inhabitants in the village when I visited it" (Cardinall 1927, p. 83).

Women's unhampered access to the labor of unfree persons—free of the hierarchical structures of household and community—has perhaps usually been restricted, even the limited availability of slave and pawn labor remind us that women are not inevitably destined to manage enterprises and perform domestic and child-care tasks on their own or with the limited assistance of children. The life of the contemporary European housewife is thus not historically inevitable.

However, I believe one of the long term consequences of the abolition of slavery as it has affected both household and women's enterprises has been a slow "squeeze" on the extent of the freedom of wives, who had hitherto had some access to subordinate labor, to dispose of their own labor time. It has led to the renegotiation of marriage arrangements as wives became either (a) the chief source of permanent labor on male controlled household farms or (b) the sole producers in female farming systems. Marriage arrangements include those between co-wives, between mothers and their children, between widows and their husbands' successors as well as between husbands and wives themselves and extend beyond the boundary of the household to the structures of labor mobilization in the community.

It is not inevitable, however, that renegotiations around marriage arrangements should leave women without labor in their own-account enterprises. Currens (1976), for example, has discussed the changes in the allocation of labor between household and private fields as a consequence of agricultural innovation in northwest Liberia. Here, men's increased opportunities to earn cash from coffee production have led them to withdraw labor from the household subsistence production of rice. This is a "classic" situation which generally has led to an increased allocation of wives' labor to subsistence production and a reduction in the time available for their own enterprises. In this case, however, it is reported that wives have been able to negotiate with their husbands to use income from coffee production to pay for wage labor to work on the household rice fields. This has enabled at least some women to maintain their labor input and even increase investment in their own private fields.

Conclusion

Most rural women's contemporary own-account enterprises—whether they be private farms, craft production or petty trading—are on a very small scale with very little assistance other than that of children and are carried on between the unreciprocated demands on her labor for household or male enterprises. The "naturalism" model attributes this to the conditions of child-bearing, child-rearing and domestic chores. It is sometimes described as work which can be picked up or put down to accommodate these other demands. The comparison with out-work in Britain is salutory; this too has been regarded as an "opportunity" for women to combine income earning with domestic responsibilities; much of it differs radically from this benign model. When women cannot mobilize the labor of others they may work themselves to exhaustion. There are, of course, successful urban women traders operating large scale enterprises, but the particular conditions of their control of labour would be the subject of another paper.

I have suggested that this is a consequence of the reduction of labor available to women, partly as the result of the abolition of slavery, partly as a consequence of the increasing appropriation of their labor by men/husbands transforming gender relations of production within the hierarchies of household, lineage and community.

It is not inevitable that women should be increasingly disadvantaged in these ways, as the example of Liberia showed. Development programs must recognize these constraints if they have any serious intention of empowering women. On the other hand, the Liberian "solution" was the employment of wage labor. This not only implies the progression of class differentiation but suggests, too, the vulnerability of women under such circumstances. For rural women as wage laborers undoubtedly lack the social power to demand wages equivalent to those of men but are potentially as "acceptable" to women employers as they are to men. The critical issue requiring further thought is this. What kind of development will take into account the pervasive structures of gender hierarchy in household and community—which include the relationships between women and women as well as between women and men—and will not disadvantage most women all of the time and some women some of the time?

References

Arhin, L. (1972) "The Ashanti Rubber Trade with the Gold Coast in the Eighteen-nineties," *Africa* 42,1.

Baumann, H. (1928) "The division of work according to sex in African hoe culture," *Africa* 1.

Beneria, L. and G. Sen (1981) "Accumulation, Reproduction, and Women's Role in Economic Development: Boserup Revisited," *Signs* 7,2.

Berry, S. (1985) *Fathers Work for their Sons*. Los Angeles: University of California Press.

Boserup, E. (1970) *Women's Role in Economic Development*. London: St. Martins.

Cardinall, A.W. (1927) *In Ashanti and Beyond*, London: Seeley, Service and Co.

Clarke, J. (1978) "Agricultural Production in a rural Yoruba community," Ph.D. thesis, University of London.

Currens, G. (1976) "Women, Men and Rice: Agricultural Innovation in Northwestern Liberia," *Human Organisation* 35, 4.

Dey, J. (1980) "Women and Rice in the Gambia," unpublished Ph.D. University of Reading.

Goody, J. (1976) *Production and Reproducton.* Cambridge: Cambridge University Press.

Guyer, J. (1984a) "Naturalism in Models of African Production," *Man* 19,4.

———. (1984b) *Family and Farm in Southern Cameroon.* African Research Studies No. 15. Boston: Boston University.

Hirschon, R., ed. (1984) *Women and Property, Women as Property.* London: Croom Helm.

Klein, Norman (1981) "The Two Asantes: Competing Interpretations of 'Slavery' in Akan-Asante Culture and Society," in P. Lovejoy, ed., *The Ideology of Slavery in Africa.* Beverly Hills: Sage.

Lubeck, P. (1985) "Islamic protest under semi-industrial capitalism: 'Yan Tatsine' explained'," *Africa,* 55, 4.

MacCormack, C. (1983) "Slaves, Slave Owners and Slave Dealers: Sherbro Coast and Hinterland," in C. Robertson and M. Klein, eds. *Women and Slavery in Africa.* Madison: University of Wisconsin Press.

Meillassoux, C. (1975) *Maidens, Meal and Money.* Cambridge: Cambridge University Press.

Mikell, G. (1984) "Filiaton, Economic Crisis and the Status of Women in Rural Ghana," *Canadian Journal of African Studies* 18,1.

Okali, C. (1983) *Cocoa and Kinship in Ghana.* London: International African Institute, Kegan Paul.

Oppong, C., C. Okali, and B. Houghton (1975) "Women Power: Retrograde Steps in Ghana," *African Studies Review* 18, 3.

Perrot, C.H. *Les Avi-Indenie et le Pouvoir aux 18eme et 19eme siecles.* Paris: Productions de la Sorbonne.

Raynaut, C. (1968) "Aspects socio-economiques de la preparation et de la circulation de la nourriture dans un village Hausa (Niger)," *Cahiers d'Etudes Africaines* 17.

Richard, P. (1983) "Farming systems and agrarian change in West Africa," *Progress in Human Geography* 7,1.

Roberts, P. (1977) "The Integration of Women into the Development Process," *IDS Bulletin* 10,3. Sussex: Institute of Development Studies.

———. (1984) "The Sexual Politics of Labor and the Household in Africa," in J. Guyer, ed., *Conceptualising the Household: Issues of Theory, Method and Application.* Cambridge, Massachussetts: Harvard University.

———. (1986) "Rural Women in Western Nigeria and Hausa Niger," in K. Young et al, eds., *Serving Two Masters.*

———. (1987) "The State and the Regulation of Marriage: Sefwi Wiawso (Ghana) 1900-1940," in H. Afshar, ed., *Women, State and Ideology.* London: Macmillan.

Robertson, C. (1983) "Post-Proclamation Slavery in Accra: A Female Affair?" in C. Robertson and M. Klein, eds., *Women and Slavery in Africa.*

Robertson, C. and M.A. Klein, eds. (1983) *Women and Slavery in Africa*. Madison: University of Wisconsin Press.

Vallenga, D.D. (1977) "Differentiation among women farmers in Two Rural Areas of Ghana," *Labor and Society* 2,2.

Whitehead, A. (1984) "Men and Women, Kinship and Property," in R. Hirschon, ed., *Women and Property, Women as Property*. London: Croom Helm.

6

Sexuality and Power on the Zambian Copperbelt: 1926–1964

Jane L. Parpart

During the precolonial period, the people of Zambia regulated sexual relations through various customs and laws. Whether in matrilineal or patrilineal societies, these customs were primarily designed to ensure male elders' control over women's productive and reproductive labor.

Initially, the colonial intrusion damaged this system by providing opportunities for men and women to escape from rural elders. Some women grabbed the chance for independence, and moved to the urban areas where they survived by engaging in various economic activities. Playing on their scarcity, women soon learned to bargain with male partners; changing partners became an accepted way to improve one's living standards.

Both African and colonial officials soon reacted in horror to the new liberated African woman. Colonial officials began to realize that their system of indirect rule would never succeed if rural chiefs lost control over women. As a result, African chiefs and colonial officials formed a patriarchal coalition and set about creating state and ideological structures to bring these women under control. To do this, they created urban African courts and new "customary" laws which redefined sexuality in terms of patriarchal power (Chanock 1985; Ault 1983; Wright 1982).

The question at issue in this paper is women's response to these pressures. Were they able to resist patriarchal domination, and if so how? Did their subjugation change over time? These questions are important not only

This chapter originally appeared in *Discovering the African Past: Studies in Honor of Daniel F. McCall*, Norman R. Bennett, ed., Boston University Papers on Africa, VIII (Boston: Boston University, African Studies Center, 1987) and is reprinted with permission.

because we need to know more about the assertion of patriarchal domination, but equally because we need to know more about the ways in which women resisted that domination and carved out areas in which they could shape their own history, albeit rarely in conditions entirely of their own choosing.

Marriage and Divorce in Precolonial Zambia

Whether matrilineal or patrilineal, precolonial Zambian societies were dominated by male elders, who maintained this dominance largely through control over marriage and inheritance. In matrilineal groups, which predominate in Zambia, property is inherited through the female line. Bridewealth payments were low, as they gave the husband no rights over children. Bridegrooms provided labor rather than cash, particularly among the labor-starved Bemba.[1] Full rights as a husband, including the right to move one's family, were won only through several years of service, gift giving and ritual acts. Wives and husbands had certain obligations to each other, but the husband never gained control over his wife's or children's property. Power and property passed from uncle to nephew rather than from father to son. Not surprisingly, relations between husbands and brother-in-laws were often tense, especially when powerful fathers tried to gain their sons' and nephews' loyalty (Richards 1939, pp. 103, 124–127). In other matrilineal groups, such as the Tonga, the sexes were more equal, and women had more control over their children's labor. The crucial tie within the homestead was between husband and wife, but tension between fathers and uncles persisted (Colson 1958, pp. 61, 137).

In all matrilineal societies, marital stability was threatened by the conflict of interest between fathers and their wives' and childrens' matrikin. Various rules were designed to contain marital discord. A wife's adultery was severely punished, especially if the "injured" husband was important. Adultery by the husband during a wife's pregnancy was believed to cause stillbirths and death. Among the Bemba, ritual ceremonies discouraged, but did not prohibit, men's adultery. And although polygyny was accepted, a monopoly on important rituals protected the first wife from younger rivals. But despite these pressures, some divorce occured in the precolonial period (Richards 1940, p. 34; Colson 1958, pp. 164–168, 176).

In Zambian patrilineal societies, a large brideprice (*lobola*) in cattle usually cemented a fathers' rights to his children in the precolonial period. Power and property passed from father to son, and patriarchal power reigned supreme. Among the Ngoni, the largest patrilineal group in Zambia, the brideprice purchased exclusive sexual rights to the wife and rights to all her children (genetrical rights). The Ngoni expected virgin brides, and obedient and faithful wives. Adultery was severely punished, usually with death, unless a chief pardoned a first offender or the injured husband could be placated with some cattle (Mitchell 1957, pp. 3-4; Barnes 1951, pp. 2-6).

Divorce was rare and could only be secured by a man; women had almost no grounds for divorce. A man could divorce his wife by sending her back to her kinfolk and giving her a small present as an indication that he had divorced her. If the wife was in the wrong, her parents could be forced to return the brideprice. Needless to say, considerable familial pressure kept most women in line (Barnes 1951, pp. 4, 119).

Among the Lozi and some Zambian Ngoni, descent is traced through either the father's or the mother's lineage. The bilateral descent system emphasizes sexual (sponsorial) rights rather than parental lineage. As a result, adulterers could claim their children. Early travellers reported widespread seduction, adultery and abduction of wives. One man acknowledged that "should a man take a liking to someone else's wife he will have an interview with her and bring her home." Marital stability depended solely on the quality of the marital relationship, which seems to have been quite brittle (Gluckman 1950, pp. 181-182).

Early Colonial Period: The Weakening of Patriarchy

When the British first arrived in Central Africa, they reacted negatively to many African customs, particularly those that apparently mistreated women. Colonial officials and missionaries indignantly set about stopping "repugnant" traditions such as inheritance of widows, forced marriage and child marriage. In order to improve their position, women were given jural status, and the right to pursue litigation at the Boma (government).

African women quickly adapted to these new opportunities, and a flood of litigation deluged colonial and chiefly officials. In the Chewa/Ngoni area of Eastern Zambia, the recognition of women's jural rights drastically altered traditional divorce law; for the first time women could divorce unsatisfactory husbands. And they did—in the early 1900s, district officials were swamped with unhappy wives bent on divorce (Chanock 1985, p. 173). Divorce rates rose among the Bemba as well, and as divorces became easier to attain, Bemba women came to rely more heavily on the support of their maternal relatives, who generally welcomed them with open arms (Gouldsbury and Sheane 1911; pp. 158, 168). In other parts of Zambia, the same story prevailed. Nyakyusa elders, for example, complained that divorce rates began increasing at the beginning of the colonial period and had increased steadily since then (Wilson 1977, pp. 190-192).

This pattern continued into the middle of the colonial period. In the 1930s, the anthropologist Audrey Richards discovered that Bemba women frequently pleaded their own case before the Boma, and were "certainly able to break a marriage contract with much greater ease than women in patrilineal Bantu societies" (1956, p. 49). Among the Lozi, already unstable marriage patterns grew more so. In 1918–20 the paramount bowed to the inevitable, and agreed to let adultery, divorce and abduction cases come before his court. They soon dominated the court agenda (Gluckman 1950, p. 181). But lineage systems gradually became less important as the divorce rates increased everywhere. By the 1940s, Ngoni divorce rates were almost as high as those among the matrilineal Lamba and Bemba peoples (Barnes 1950, pp. 50–51).

But the greatest changes for women came as a result of the penetration of the colonial capitalist economy. Colonial policy pushed men into migrant labor, leaving women stranded in the rural areas with an increasingly onerous work load. As rural conditions deteriorated, the cities beckoned. While women had little chance for waged employment in town, other opportunities to earn money existed. Beer brewing, gardening, selling food and services (including one's body), and above all, partnerships with men, offered women the means to survive in town (Chauncey 1981, p. 159). After 1926 the copper mining companies[2] facilitated female migration by encouraging more skilled miners to bring their families to the mines during labor contracts (Parpart 1986a).

Despite opposition from rural chiefs, women migrated to town in large numbers. While statistics on female migration in Zambia are inadequate, we do know that by 1931, about 30 percent (or 5292) of the 15,876 black mine employees lived with their wives on the Copperbelt, and more women lived in the nearby government townships. By the 1940s, about 15,000 women lived on the mines (Parpart 1986a, pp. 142–143; Perrings 1979, p. 252). In 1955, the Copperbelt mine townships had a population of 44,682 men, 29,146 women and 71,801 children, while the municipal townships had 32,443 men, 15,575 women, and 24,111 children (Passmore 1956). By 1961, about 80 percent of the black miners had wives at the mines (Parpart 1986a, p. 143). Sex ratios continued to even out, and in 1969, the urban Copperbelt area had a population of 205,117 men and 166,394 women (Zambia, 1969 Census).

Once in town most women found ways of earning income, but very few became self-sufficient. Most women needed some extra support, which usually came from men. The potential problems of such dependence—vulnerability to cruel or niggardly mates—were mitigated by favourable sex ratios. In 1939 men outnumbered women 2:1,[3] but in 1954 the ratio was still 169 men to every 100 women (Mitchell 1961, p. 8). "Pick-up" marriages became the norm on the Copperbelt. In nearby Broken Hill,[4] Godfrey Wilson discovered that "The younger married women all have alternative mates readily available, and this abnormal fact reduces the disadvantages of divorce for them, though not for their husbands. Little domestic disputes and incompatibilities, therefore . . . now lead many women to leave their husbands" (Wilson 1940, II, p. 65). The Copperbelt was much the same (Ault 1983, pp. 182–87; Chanock 1985, pp. 206–8). As the Secretary for Native Affairs wrote in 1936, "The 'mine marriage' has become notorious and natives have told me here on the railway line that marriage according to native law and custom does not exist . . . [One] money making method is to contract a mine marriage and to get as much out of a husband as soon as possible for about two months. The woman then marches out of the hut and marries someone else" (Epstein 1953a, p. 59).

Copperbelt women continued to pursue marital cases in court as well. In 1936, one district commissioner admitted he dealt with so many mine marriages that he had come to view all matrimonial disputes with suspicion (Chanock 1985, p. 206). Thus by the middle of the colonial period, women were demanding their rights within marriages and asserting their right to break off unsatisfactory marriages either informally or through the courts.

The Reassertion of Patriarchal Power

The growing autonomy of women in rural and especially in urban areas began to disturb both African and colonial authorities. By 1915, at least one colonial administrator took the view that loosening the ties of African matrimony had been a mistake. "We have freely granted divorces in favour of frivolous girls, and permitted them to run from one man to another" (Dundas 1921, pp. 263–266). African chiefs had equally damning things to say about independent women (Chanock 1985, p. 192). Colonial and rural African officials thus cast about for ways to reassert patriarchal power. They recognized the connection between the control over sexual behavior and the authority of the chiefs, and perceived women's new found freedom as a threat to that authority, and consequently a threat to the system of indirect rule.

The field of marriage, adultery and divorce was seen as a crucial arena where defeated African authorities could reassert their power. To that end, Native Authorities were set up in 1929. They were given judicial powers and charged with the responsibility of establishing law and order in the rural areas. The Native Authorities set about creating a new customary law that expanded chiefly powers and brought "frivolous" women under control through the regulation of newly amended "traditional" marriage, divorce, child custody and inheritance laws (Chanock 1985, ch. 8).

The native courts began to insist on registered marriages, some form of brideprice payment, and the treatment of adultery as a civil offence with severe financial penalties. Registration involved the consent of parents or guardians and the native authorities; a woman's consent became less important. Thus marriage certificates reinforced the role of the family and enlarged the powers of the chief. Matrilineal leaders worried that matrikin would lose out, but in the end the chance to increase control over headstrong women outweighed these reservations. Chiefs realized that marriage certificates could be used to control women's movements to town. In addition women in town without certificates could be harrassed and even repatriated to the rural areas. The native authorities also favored registration because it would reduce intertribal marriage and reassert chiefly control over over impudent young migrant laborers with their pay checks and their "newfangled ideas" about marriage and seniority (Chanock 1985, ch. 10).

Even native authorities without a strong tradition of bridewealth payments began to encourage higher payments to "secure" marriages.[5] Higher payments reduced a woman's ability to leave a marriage, as her family had to repay the brideprice at divorce unless it was the husband's fault. This trend reflected a concern by tribal elders with controlling wives and children, and a desire to keep young male migrants tied to them through the need to acquire large payments. But it also reflected migrants' preference for cash rather than labor payments to in-laws, and a growing desire to keep wealth within the nuclear family. Men wanted both to pass on their accumulated wealth to their children and to gain rights to their children's future earnings (Chanock 1985, pp. 178–181; Mitchell 1957, p. 27).

The Native Authorities' third strategy for the control of women was to punish adultery, particularly when it involved abduction. As we have seen, during the early colonial period changing partners had become the norm on the Copperbelt. African and colonial officials disliked this behavior, seeing it as a symptom of moral decline and female indiscipline. In order to contain it, traditional leaders created a new customary law based on a largely fanciful reinterpretation of traditional African law. Adultery and abduction were declared serious civil offences, and large compensation payments (3 to 7 pounds) were awarded to injured husbands. Women had to pay a smaller fine as well. The courts sought to discourage making a business out of adultery by lowering fines for repeat offenders. Registration was encouraged by the court's refusal to award compensation to husbands in unregistered marriages (Chanock 1985, ch. 11; Epstein 1981, pp. 315–318).

Abductors were dealt with particularly harshly, as they were punished for the much more heinous crime of destroying a marriage. They often had to pay compensation of 7 pounds or more, and received severe reprimands from court authorities (Epstein court records).[6]

In the mid 1930s, these regulations were largely ineffective in towns since women preferred the flexibility of informal liasons. The mining companies' disinterest in strict registration procedures for married workers did not help. As the Mufulira District Officer regretfully reported in 1945, the compound managers merely want a statement that a couple are "probably in a genuine marriage and deserve housing." As long as one wife at a time lived in married housing, management was content.[7] The mining companies wanted contented workers, not bureaucratic hassles.

In contrast, colonial administrators, missionaries, and both rural and urban African leaders became increasingly alarmed about the informal marital arrangements in town. They warned of a rising tide of urban immorality, crime, and social disorder—prophesies that seemed all too true when the Copperbelt-wide 1935 strike brought rioting and death. Both African and colonial authorities blamed some of these problems on independent immoral women and agreed that these women must be brought under male control. To accomplish this and to strengthen the influence of Native Authorities in town, the colonial government established urban African courts in the major urban centers between 1936 and 1939. The Native Authorities appointed urban court members (assessors) and charged them with enforcing the new "customary" law developing in the countryside. As in the rural areas, the goal of improving marital stability and controlling wayward women was high on the agenda (Epstein 1953).

Some tension existed between the urban and rural courts over jurisdiction, but both groups believed the courts should stabilize urban marriages. Initially registration of marriage and divorce had to be done through the native authorities, but urban couples protested and by the late 1940s, urban courts were marrying and divorcing urban-based couples. In 1949, five Copperbelt urban courts dealt with 1148 divorce cases, which accounted for 33.62 percent of the year's civil cases (Epstein 1953a, p. 52). And these figures underestimate divorce activity because couples whose marriages were contracted in the rural areas were usually sent home for divorce. And while urban court members sometimes disagreed with native authorities, they supported their rural counterparts' desire to control women. The urban courts insisted on parental permission for marriage, proper marital registration, and high brideprice payments. They discouraged intertribal marriage, only reluctantly granted divorces, and handed out heavy fines to adulterers and abductors (Epstein 1958, ch. 3). Despite repeated failures, the courts also doggedly continued trying to send unmarried women home to the rural areas.[8]

As marriage certificates became more necessary in town, and as men realised they could control wives and benefit from women's sexual "misbehavior" better in properly registered marriages, the urban courts gained increased power over people's private lives. More people entered registered marriages, resulting in more litigation. Indeed, by 1949 seventy percent of urban court cases involved matrimonial issues of one kind or another (Epstein 1953a, p. 8ff.).

Tribal elders on the Copperbelt also adjudicated less serious cases. Established in all the townships by 1940, the elders offered a more informal means of mediating problems along customary lines. The elders were urban residents, but usually older and of high status, often well connected to rural leaders. They were elected by their urban tribesmen, but had moral rather than judicial power. They attracted people wanting advice rather than punishment. Severe cases were referred to the urban court. The elders resented the court's judicial authority, which inevitably undermined their own. Their position was made worse by the new African Mine Workers' Union, which voted the elders out of the mine townships in 1953, leaving them only in the government townships. Nevertheless, these elders continued to provide an informal court for urban Africans, and much of their time was spent adjudicating marital squabbles. They condemned casual marriages and independent women. And while resentful of the courts, they too saw themselves as the guardians of marital stability and patriarchal authority on the Copperbelt (Epstein 1959, pp. 48–60).

As a result of these efforts, marriages stabilized somewhat on the Copperbelt. As early as 1943, Labor Department researcher Lynn Saffery discovered a fair degree of stability among African marriages (1943, p. 41). In the early 1950s, the Rhodes-Livingstone researchers reported fairly equal degrees of stability between urban and rural African marriages. In 1952, Clyde Mitchell evaluated 430 marriages at Nkana mine township. Of these, divorce had dissolved 22.1 percent of the marriages contracted in the rural areas and 25.6 percent of the urban marriages. He contrasted these divorce rates with those in the rural areas (the Yao, 41.3 percent; the Ngoni, 36.9 percent, and the Lamba, 41.8 percent), and concluded that urban marriages were not noticeably less stable than their rural counterparts (Mitchell 1957, p. 10). While admitting in a later article that urban first marriages divorce more than rural first marriages (47.1 percent would survive 20 years as opposed to 68.9 percent of rural marriages), Mitchell never denied his earlier conclusion that urbanization had not significantly altered Zambia's divorce rate (Mitchell 1963, pp. 260–261). And this, he maintained, was largely due to the urban courts which "by explicitly stating, in their judgments, the norms they consider appropriate and by punishing deviance from these norms, are gradually bringing about a type of marriage which is independent of particular tribal custom and consonant with town living" (Mitchell 1957, p. 29).

Urban Women Fight Back

The clamp down on sexual freedom, along with restrictions on beer-brewing and prostitution, limited women's opportunities to survive independently on the Copperbelt. The literature has focussed on the closure of economic opportunities for urban women, particularly restrictions on changing partners. As James Ault concluded, Zambian women may have breathed the free air of the city in the 1930s, but that was no longer true by the 1950s (1983, p. 192). Marriage had been "traditionalized" in the urban centers and most women had been brought reluctantly into line.

While fundamentally correct, this line of argument presents women as passive pawns in patriarchal struggles. Men seem to win an easy victory, with women readily accepting limitations on the freedom so fully enjoyed in the 1920s and 1930s. This is too easy a transition. The urban court records of the 1950s suggest another scenario, one where women could and did assert some power in a changing environment.

As patriarchal forces increasingly constricted women in the urban areas, women fought back in a number of ways. Some sought to avoid male control by remaining economically independent. A few women had "respectable" jobs such as nurses, teachers and welfare assistants, but waged work for women was rare. Unlike West Africa, women did not dominate the marketplace in Zambia (Brelsford 1947). Most independent women made a living from beer brewing and prostitution, but the disreputable and illegal nature of these livelihoods made them an easy target for the state (Epstein 1981, pp. 116, 309–310; Chauncey 1981). Court members viewed independent women as harlots, and felt no compunction about fining them. Indeed some Copperbelt court members admitted that "the courts do from time to time [simply] impose fines on unattached women visibly existing without means of support."[9] Women who lived alone on the Copperbelt received little sympathy in the courts. In a typical example, court members disapprovingly told a woman litigant that "you can not be a good woman, otherwise you could not have lived on the Copperbelt for nine years without getting married."[10] Whenever possible, the courts repatriated such women in an effort to bring them under the control of rural (i.e. male) authorities.

Women did not accept this treatment without protest. Repatriated women reappeared in town in such numbers that urban court members abandoned the policy of female repatriation in 1953.[11] But women increasingly recognized the importance of male partners in town, both for economic support and legal protection. By 1954, only 8 percent of Copperbelt women lived alone— 2 percent were divorcees or widows, five percent had never married, and one percent lived apart from husbands. Ninety-two percent lived with a man, in either legal or informal arrangements. This is not surprising considering official sanctions against single women and continuing favorable sex ratios (Mitchell 1957, p. 8).

But pushing urban women into liasons did not necessarily control them. Informal relationships were difficult to monitor, and many women preferred them for that reason. The urban courts refused to consider most marital disputes between unregistered couples. As a result, women in informal marriages could leave unsatisfactory partners without facing a hostile urban court. Unregistered husbands could neither stop a woman from leaving nor sue her lover for compensation. Women in intertribal marriages in particular preferred unregistered marriages in order to avoid court battles over residence and child custody.[12]

While inadequate data impede quantification, unregistered marriages apparently continued to flourish, especially among poorer Africans. In 1943 colonial officials reported widespread resistance to marriage registration among "the bulk of natives living near the centres of employment . . . the main objection to registration comes from the least responsible elements . . . and is based on the fear that once a marriage certificate is taken out the marriage will be more binding then they wish. The women are particularly inclined to hold this view."[13] Registration increased after 1944, but informal liasons and resentment against registration continued to flourish. In 1954, women on the Copperbelt vociferously opposed marriage registration during a demonstration against the colonial regime.[14] And as late as 1964, mine managers admitted that many employees simply changed wives without informing company officials.[15]

Women also evaded male control by avoiding brideprice payments. A high brideprice was supposed to make wives less "proud and cheeky" (Mitchell 1957, p. 25). As in the case of marriage registration, the courts refused to protect husbands who had not paid a brideprice. Consequently, some women preferred looser arrangements. As one woman told the urban courts, once her people returned the brideprice to her ex-husband, "she stayed on with him, saying that since the bride price had been paid she could easily leave him if she wanted."[16]

Even in registered marriages, the families of women from matrilineal groups generally preferred smaller brideprice payments. Mitchell discovered that among 172 couples with matrilineal brides, 43 percent paid the lowest price, while only 18 percent paid the highest. In contrast, couples with brides from patrilineal societies were absent from the lowest group while 60 percent paid the highest fee (Mitchell 1957, p. 23). These lower payments ensured the rights of a husband to his wife's body and labor, but not to her progeny. It is probably safe to assume that larger payments to matrilineal kin were paid by prosperous husbands wishing to ensure control over their children.

But some wives and their families refused to accept brideprice payments at all. This seems to have been most common in marriages between matrilineal women and patrilineal men, with their inevitable disputes over child custody. Although the urban courts usually awarded custody based on the customary law of the mother, a high brideprice could be used to argue for a father's right to custody. In one case, a Bemba woman fought her husband in four courts before winning custody.[17] In another, a prominent Ngoni man wrested custody of his sons from his Bemba ex-wife by convincing the courts that he could better ensure their future.[18] Such cases concerned reluctant in-laws, and some parents from matrilineal societies refused the brideprice rather than lose control over their grandchildren. In one case a man complained that his in-laws refused the proferred payment. They told him, "Don't worry us. You're just a temporary husband. We want somebody from home." Later they persuaded their daughter to obtain a divorce.[19]

However, as pressures for properly registered marriages and greater marital stability increased in the 1940s and 1950s, more urban couples needed and acquired legitimate marriage certificates. As a result, the common solution to an unsatisfactory relationship, namely changing partners, became more difficult, especially for women. Husbands in properly registered marriages could sue a wife's lover for large compensation payments, making it difficult to change partners without first getting divorced. Husbands could block divorces as well. Courts disliked awarding divorces, particularly if the husband wanted the marriage to continue. In one case, for example, a wife demanded a divorce because her husband "troubled her, and told her to leave the house. . . . All he was after was the household property." But the husband opposed a divorce. The court members ignored the wife, listened sympathetically to the husband's testimony, and decided that "The best thing is that we should attempt to instruct your husband how he should look after his wife." While temporarily taken aback by the wife's vehement rejection of their decision, the court reaffirmed its decision after hearing unsympathetic testimony from her grandfather and learning that this was her third marriage.[20]

But litigation was a two-edged sword, and some women succeeded in using the courts and other fora to their advantage. They learned the value of protest, and the need to frame arguments in certain ways. Since men were permitted sexual access to more than one woman, women could not sue men for adultery or polygyny. But colonial and African authorities asserted certain moral values that supplied women with grounds for litigation. According to these norms, "proper marriages" were stable; a good wife bore children, prepared food, cleaned house, and remained sexually faithful to her husband. But in return, husbands were supposed to be generous, kind and responsible. Women played on these values by complaining to the authorities about neglect, assault, and disease, rather than male philandering and polygyny. While the latter often lay behind formal accusations, women quickly learned to argue cases on grounds they could win.

As the urban courts gained prestige and authority, they became increasingly important for the solution of serious, and even not so serious disputes. But a number of intermediate fora also existed. In the municipal townships, elected tribal representatives provided both registered and unregistered couples with important alternatives for quarrels that couldn't be solved at home. These elders invoked chiefly authority through traditional law and their decisions carried considerable moral force. Although unable to inflict punishments, the elders expected their decisions to be binding. Repeat offenders were sharply rebuked. The elders had a high success rate in matrimonial cases. In 1963, the Mikomfwa[21] elders reported at least temporary reconciliations for 67 percent of their 131 matrimonial cases. Only 33 percent were referred to urban court. Rivalries with urban courts probably kept the number of referrals down, as the elders liked to think of themselves as the more authentic fount of knowledge about traditional law (Harries-Jones 1964, pp. 33–34, 64–65).

Disgruntled couples took matrimonial disputes to government welfare officers as well, but their reconciliation rate differed dramatically. In 1963, for example, the Luanshya government welfare officer saw 92 cases, yet referred 69 percent of them to the urban courts. The welfare officers' close association with the urban court gave people the impression that a visit to the welfare officer was just a time-consuming impediment on the way to court (Harries-Jones 1964, pp. 64–65).

Before 1953 the elder system in the mine compounds performed much the same function as it did in the government townships. After the TRs were abolished, corporate social welfare officers tried to fill the breach. Compound managers had long acted as an informal appeals court to handle minor domestic quarrels, but after 1953, the copper companies' growing commitment to stabilized African labor induced management to upgrade its social welfare facilities. Mine management invested heavily in trained case workers who spent considerable time dealing with workers' domestic disputes. In 1959, Roan set up a Citizen's Advice Bureau run by company case workers. Like the elders, the CAB was an advisory body, although it could invoke sanctions from management. It dealt primarily with marital problems, usually brought to them by women between 21 and 27 years, in the first year of their marriages to largely unskilled older men. In six months (1960–1961), fully 83.8 percent of the 538 cases were initiated by women. They accused their husbands of neglect, desertion and assault, often in connection with beer drinking and womanizing. About 18 percent of the cases involved polygynous unions—an unusually high proportion considering only 2.4 percent of the mine population lived in polygynous households. While serious cases were referred to urban courts, temporary reconciliation rates were high—about 49 percent, moving to 86 percent when the Bureau moved out of a building identified with the Roan sub-court (Harries-Jones 1964, pp. 35, 44–46, 64–65). But the nature and frequency of these largely female complaints certainly undermines any assumption that women passively complied with increasing patriarchal authority in the home and the community.

Women in the Urban Courts

Neither tribal elders nor social case workers had the authority to enforce discipline or punish offenders, so increasingly Africans took serious quarrels to the urban courts. In these courts women expressed some of their most determined and creative opposition to patriarchal domination.

Women seem to have initiated more of the marital cases brought before urban courts, while men usually just sued their wives for adultery. In 1950 out of 85 adultery cases in the urban courts of Ndola and Broken Hill, only one was brought by a woman. In contrast, seventy-five percent of the non-adulterous marital cases (42 out of 56) were initiated by the wife, and three were jointly initiated. In Mufulira, there were 45 divorce cases in three months during 1951: 25 brought by women and 20 by men (EP: court records 1950–51). Similar patterns also emerged in the rural courts. In 1963, Ndola rural native authority dealt with 219 cases; 88 percent were brought by women who charged their husbands with: neglect (38 percent), assault (26 percent), and desertion (10 percent). Husbands accused wives on two counts: adultery and disobedience, and these took up only 6.8 and 5.4 percent of the cases respectively. This preponderance of female plaintiffs suggests greater discontent among women. However, it is possible that since men seeking divorce with insufficient cause were treated harshly by the courts, some men may have pushed their wives into court instead (Harries-Jones 1964, pp. 45, 48).

As noted, women quickly learned to avoid complaining in court about adultery or polygyny. Occasionally a wife took her husband to court for adultery, but usually to no avail. One wife demanded a divorce because her husband "likes adultery as if it is the work he does." While the court sternly upbraided the husband for his behavior, they refused the divorce.[22] Complaints about polygyny received similar reactions, although urban court members generally discouraged polygyny in the urban areas for all but the very rich.[23]

In order to win a case against one's husband, women had to couch their complaints in terms of neglect, desertion, assaults and disease—all punishable behavior as far as the courts were concerned. Neglect was the most common grounds for divorce, and women discovered the advantage of presenting themselves as good faithful wives, victimized by neglectful husbands. In a typical case, the wife alleged that "her husband was always troubling and worrying her. They had been before the court earlier when the woman had been involved in an adultery case. But thereafter the husband had not cared for her, nor provided her with clothes, and food." She accused him of being a drunkard. "When he gets his pay he can't sleep in the house, and he does not regard me as his wife."[24] Women frequently complained of being "chased from the house" as well—a common occurrence on the Copperbelt where housing was tied to jobs. As one woman told a sympathetic court, her husband had told her "not to come to the house any more, he did not want her, and if she came he would kill her."[25] When neglect and abuse could be proven, urban courts were more apt to grant a divorce.

And even when they refused, a district officer could overturn the decision if convinced of the woman's case.[26]

A husband's desertion could be grounds for divorce as well, though this reason featured more prominently in rural courts which dealt with abandoned migrant laborers' wives. In the urban areas, accusations of desertion were usually coupled with neglect. For example, a woman was awarded a divorce because "while she had been ill for seven months her husband had not come to her, and she had had to stay with her father." Furthermore, she insisted that her husband had never given her enough food during their two year marriage.[27]

Women could also divorce a man for impotence. Wives were not expected to have to remain with impotent husbands. And while the court preferred such cases to be solved within the family, if that was impossible, they usually granted a divorce.[28]

While polygyny was never grounds for divorce, an unusually high percentage of cases came from polygynous households.[29] Polygyny exacerbated conflicts within the family. One wife, for example, burned her husband's trousers and all her certificates when she discovered his new wife.[30] Another won a divorce because her husband refused to sleep with her while the other wife was pregnant.[31] Even wealthy men were not immune from such cases. One court assessor's wife brought him to court, claiming that "Although they had been married long, and she had borne him five children, she was given only her old clothes to wear while he gave his younger wives money so that they could go round in new ones. She demanded a divorce." She eventually won her case.[32] And in general, urban court members listened to these cases with a more sympathetic ear because while recognizing polygyny's legality, for the most part they disapproved of it in the urban areas.[33]

Divorce cases often indirectly indited unfaithful or inattentive husbands as well. In a typical case, a wife claimed neglect, but it was more emotional than physical. The husband spent his free time with a girlfriend, and ignored his pregnant wife. He refused to mend his ways, and exasperated urban court members agreed to award a divorce after the birth of the child.[34]

But presenting winnable grounds was not enough. Women discovered other means of winning divorce cases as well. Probably the most effective was strong support from parents or guardians. This impressed court members, particularly if the support came from a senior male, and if the marriage was intertribal or of short duration. For example, in an appeals case a husband vehemently denied his wife's accusations of frequent beatings, and demanded that the marriage continue. He argued that "Of course I am bound to beat my wife if she does not behave well, but this would not mean dissolving of the marriage." But the court members listened to the wife and her parents who insisted on a divorce. The court chastized the husband, telling him that "the parents of the girl agree that you must divorce. We have nothing to do with this. Once your parents-in-law do not want this marriage it is finished. We are bound to agree with them."[35]

Even when judgments went against in-laws, their grievances were often taken seriously by the court members. In one case, a son-in-law had rejected his father-in-law's attempt to return the brideprice payment and accused him of trying to destroy his marriage. The court refused the divorce, partly no doubt because other relatives and the wife wanted the marriage to continue. However, the court warned the young man that "he had been guilty of disrespect towards his parents-in-law . . . , in future he must learn to give them proper respect and not go about saying bad things against them."[36]

Supportive relatives could help women obtain reasonable property settlements as well. The courts preferred to divide marital property equally, but decisions varied. Supportive relatives could sway the court's decision, especially over brideprice repayments. Angry husbands often demanded excessive repayments for expenses incurred during a marriage. A stern rebuttal from parents or guardians was necessary at that point. In one such case, the husband wanted a repayment of 27 pounds, but only received the L3-2s-6 that his in-laws agreed to.[37]

There were other ways to win divorces from reluctant husbands. Some women just kept dragging their husbands before elders and courts until the husbands agreed to a divorce. In one case, the woman was granted a divorce on her third application.[38] Determined wives sometimes returned to the rural areas to get a divorce, no doubt assuming loyal relatives could help win the case.[39]

Women discovered that repeated adulteries could drive reluctant husbands to divorce, and that the courts would encourage such decisions. One woman was upbraided by the court for committing adultery with other men. They were particularly horrified "that she stopped her child to suck from her because she wanted to commit adultery with other men," and advised the reluctant husband to accept a divorce.[40] In another case, the court advised a husband to divorce his philandering wife after she had been brought in for causing a brawl.[41] Women also discovered they could escape arranged marriages by running around with other men. One distraught husband in such a case finally took his wife to court "to know why I am not loved by my wife." The court advised him to give up trying to win his wife's love and to get a divorce.[42] One woman even turned adultery accusations on their head by successfully suing her husband for divorce on the grounds that he forced her into repeated adulteries for his own profit.[43]

When all else failed, a public scene sometimes worked, especially with prominent husbands. A court assessor's wife sued him for divorce. After initially refusing the case, the court and the assessor relented when "On the next day she [the wife] returned to the court with the children all carrying the household goods so they could be distributed. A large crowd of spectators gathered around the court to hear the proceedings. . . . Yaka [the husband] told his fellows that they would have to hear the case themselves. It did not matter. He was disgraced. Let the matter finish." The wife remained adamant. "Eventually it was agreed that the woman should go home where she would get a certificate of divorce, and the goods were divided between them. No record was entered."[44]

Moreover, some women in registered marriages simply ignored the courts, and divorced themselves. For example, a woman in a divorce case against her third husband, explained that her first marriage "had been a 'runner,' then there had been John from whom she had 'divorced herself,' i.e. without coming to court."[45] Sometimes these informal divorces caused trouble because the legitimate husband could sue his estranged wife's lover for adultery. The court records are full of such instances, which suggest that while potential adultery cases may have inhibited informal divorces, they still remained a popular option. Indeed, according to the urban court members, certain women, particularly the Nyakyusa, "were famous for deserting and going off with other men."[46] Other women, such as the Bemba, were considered hard to handle and prone to desertion.[47]

Thus, women on the Copperbelt discovered many ways to obtain divorces despite patriarchal efforts to thwart them. As we have seen, in the early 1950s only 47 percent of Copperbelt first marriages survived twenty years, although patrilineal marriages seem to have survived better than matrilineal ones.[48] Mitchell's 1951 divorce survey concluded that 62 percent of men's marriages and 55 percent of women's marriages in Luanshya were remarriages (1963, p. 260; 1957, p. 8). But Powdermaker and Epstein believe these figures underestimate the actual divorce rate, particularly if temporary unions are included. Mitchell's study was limited to people registering their marriages in urban courts, something many couples never did. Epstein also discovered that many informants ignored temporary unions when discussing their marital histories (Powdermaker 1962, p. 161; Epstein 1981, p. 291). And while exact data are hard to come by, the capacity of dissatisfied women to obtain either legal or informal divorces was well known. As one trade union leader put it, "wives . . . had considerable influence on their husbands, and grew tired of their husbands off all the time to different meetings. They wanted their husbands to stay at home and make them happy. That was why so many of them were always getting divorced."[49]

Marital Disputes

Divorce was not the only solution to marital problems. Women found other less drastic means to improve their position within marriage. To protect themselves from divorce, many wives squirreled away private savings, often in different houses, to avoid detection. One woman reported that "her brother's wife asks her to keep money for her because she does not want him to see it. At this time I am keeping the amount of 17 shillings for her."[50] The sums were usually small, but they provided some security. Women also invested in presents for their relatives to ensure support in the event of a marital breakdown.

Many women took matrimonial disputes to court, not to obtain a divorce, but to improve their husband's behavior. Cases usually went to elders or social case workers first, but serious offenses were considered by the urban courts. Assault was a fairly common complaint. Wives took their partners to court both to win compensation and to teach them a lesson. Plaintiffs with sufficient evidence, and a "good character," usually won. The court members disapproved of male violence against women, though they believed harlots deserved harsh treatment. Convicted husbands were given a lecture on proper behavior and fined. Compensation payments could be high. One man was ordered to pay 1 pound, and after a repeat attack, another 5 pound fine was levied. This reflects the traditional belief that fines should vary with the importance of the persons involved and the nature of the crime. The more important the person and more "unnatural" the crime, the higher the fine.[51]

Women also sued husbands for inflicting sexual diseases on them, especially syphillis and gonnorhea. Often, of course, the accusation of infidelity was involved as well. Again, with sufficient evidence, the plaintiff usually won the case. Fines varied with the disease's severity, but they were usually several pounds. In one Ndola case, for example, the court fined the husband 3 pounds, a large sum for an ordinary worker.[52]

Although adultery cases were primarily used by men against women, sometimes women were able to defend themselves or to use adultery for their own end. Wrongly accused adulteresses objected strenuously, and the courts threw many adultery cases out for lack of evidence. Epstein's Copperbelt court records for 1950–51 reflect this situation: 44 (34.4 percent) of the 128 adultery cases were dismissed.[53] As we have seen, women sometimes repeatedly committed adulteries to gain a divorce. But some wives made money from adultery, both for themselves and their husbands. The "business of adultery" was so common that the courts felt the need to take extra precautions against it, such as lowering compensation payments for repeat offenders. But the very need to take such steps proves the prevalence of profit motivated adultery: obviously this was a recognized means of obtaining some ready cash.[54] One woman even took her husband to court for refusing to share the profits from her adultery.[55] Accused wives frequently excused their adultery on the grounds of neglect, in the hope that a court reprimand would improve their husbands' behavior.[56]

Sometimes wives just took their husbands to the courts or the authorities to complain about their behavior and to get advice on "how to live." In one case a wife sued her husband claiming he wanted to take another wife. He denied the charge, and the court sent them both home telling them "to live well together."[57] Another woman brought a case to court, claiming that there were troubles in the home, and "I want the urban court to help us live together." The husband admitted quarreling, but wanted the marriage to continue. The court threw the case out and told them to go home and behave responsibly towards each other.[58]

Even when a woman lost a divorce case, she usually gained some support from the court if she convinced them she was the injured party. The courts generally gave errant husbands a lecture even when denying the case. In one case, for example, a woman in a polygynous union claimed severe neglect and demanded a divorce. The courts refused the divorce, but criticized the husband, claiming that "you made a bad mistake when you bought your other wife a house leaving this one alone with her children. For that she has suffered a lot. . . . You must try and buy a house for your wife [the plaintiff]. We don't want you to neglect her for you married her at home."[59] This type of advice at least gave the wife some leverage to demand better care, and if dissatisfied, she could always return to court.

A woman's class position and her ethnic group also affected her chance of a favorable response from the courts. Elite women were under more pressure to stay in marriages, often due to high bridewealth payments and a rich husband's desire to keep wealth within the family. This was particularly true in patrilineal societies. We do not know enough about elite marriage and divorce patterns, but middle class marriages were apparently more stable than the average urban marriage (Powdermaker 1962, p. 152). Education affected divorce rates as well. Mitchell discovered that uneducated urban women had a 27 percent divorce rate, while women with a standard VI education reported a rate of only 1.8 percent (Mitchell 1986, personal communication; see also footnote no. 48). Marriage patterns of chiefs differed as well. Commiting adultery with a chief's wife or daughter brought severe penalties, particularly in patrilineal societies.[60] At the same time, daughters of the urban and rural elite could readily gain a divorce if they had family support. Chief's daughters were never exchanged with brideprice and a divorced husband could claim no return for marital expenses. The few relevant cases in Epstein's court records suggest that influential parents could sway court opinion in their favor. Of course, this only worked to women's advantage if the family backed the case,[61] and since family support usually depended on patriarchal approval, independent women often paid a price for such family support. That price was greater obedience to male authority in both patrilineal and matrilineal societies.

Conclusion

In precolonial Zambia, male elders' control over women's productive and reproductive labor was a key ingredient for maintaining patriarchal authority. Initially colonialism interrupted this system and provided opportunities for women and junior men to escape from senior male domination. New jural rights for women and emerging economic opportunities particularly in the towns, enabled men and women for the first time to survive outside their natal societies, and thus escape gerontocratic rule.

Women quickly took up these opportunities, and rural divorce rates soared as women asserted their independence. But the towns beckoned as well. To escape deteriorating rural conditions and rural patriarchs, women moved to the urban centers in increasing numbers. Most survived by a combination of economic activities and some male support. This dependence placed them in a potentially vulnerable position, but favorable sex ratios provided some leverage, and as a result informal liasons became a mechanism for survival for Copperbelt women. The notorious short-lived mine marriage soon dominated the Copperbelt.

This new found freedom for women quickly exasperated both African and colonial authorities. Missionaries and colonial administrators stopped worrying about liberating women and started worrying about controlling them. Rural chiefs expressed similar sentiments. The adoption of indirect rule in 1929 brought matters to a head, and the colonial state, rural African authorities and urban African elites set about using political, economic and ideological weapons to bring women to heel. Opportunities for economic autonomy were reduced and new "customary laws" sought to control women's sexuality and freedom of movement. Church and state authorities buttressed these laws with well developed patriarchal ideologies, which emphasized modesty and subservience as ideal female traits.

This paper does not deny the emergence of patriarchal power in colonial Zambia, but asserts the need to recognize that this power had to be won each day anew. Women fought back, often successfully. Like workers, they fought to improve the conditions under which they labored both as producers and reproducers. And like class struggles, gender struggles were mediated by political, economic and ideological factors. Poor women fought different battles, and for different rewards, than middle class women. Poor women were more interested in autonomy as they had less to gain from marriage, whereas elite wives had more to gain within marriage, so frequently used medicines, religious guidance, and other means to improve their status within the institution of marriage (Keller 1978). In contrast independently wealthy women had less to gain from marriage, and as in the Luapula region, have continued to assert matrilineal prerogatives, particularly the right to keep inheritance in the female line (Poewe 1978 (4), pp. 356–365).

Membership in ethnic and religious communities undoubtedly altered the options and constraints available to women. But whatever one's class, ethnic or religious background, women, unlike men, had to shape their lives within the constraints posed by patriarchy. The double burden of reproductive and productive labor and the male monopoly over waged employment limited women's economic opportunities and thus their ability to challenge male dominance, but it did not keep them from struggling to define and improve their lives within these constraints.

Indeed this paper leaves no doubt that Zambian women often succeeded in defending their interests even in hostile circumstances. They discovered ways to use new patriarchal institutions and laws to gain their new ends. The contest to control female sexuality was not purely one-sided. Women learned how to use their sexuality to bargain with men both in the courts and in daily life. The battle between the sexes has been waged long and hard on the Copperbelt, and while women and men have often joined forces over important issues such as strikes (Parpart 1986b), gender struggles have dominated the daily lives of women and men. The intensity of that struggle is revealed by the general paranoia which characterized male-female relationships on the Copperbelt in the 1950s (Powdermaker 1962, pp. 166–167; Epstein 1981, pp. 328), and the antagonisms and cleavages between the sexes which continue in Zambia today (Poewe 1978; Schuster 1979, ch. 8). This history serves to emphasize the importance of gender and the need to incorporate gender into the analysis of Zambian and African history.

Notes

1. The Bemba are the largest matrilineal group in Zambia and are based in the North East.

2. Two mining companies dominated the Copperbelt: Anglo American (AA) and Rhodesian Selection Trust (RST). The two major RST mines were Roan Antelope Copper Mine (RACM) and Mufulira Copper Mine (MCM). The two major AA mines were Rhokana Copper Mine (Nkana) and Nchanga Consolidated Copper Mine (NCCM). They are located in the towns of Luanshya, Mufulira, Kitwe and Chingola, respectively. Ndola is the commercial center of the Copperbelt.

3. Zambia National Archives (ZA) SEC/NAT/66G: Labour Department Annual Report, Chingola Station, 1939.

4. A city just south of the Copperbelt on the line of rail (the railroad south to Zimbabwe). Now known as Kabwe.

5. This payment (called *mpango* by the Bemba) gave men sexual rights to their partner. *Chisungu*, or the virginity payment, was not repaid at divorce.

6. A.L. Epstein's court records (designated EP) span the period 1950–55. This paper would not have been possible without Professor Epstein's kind permission to look at this material.

7. EP: D.C. Mufulira, Mr. Chicken, "Marriage Registration," February 1945.

8. EP: Minutes of the second urban court members' conference at Lusaka, 13–16 April 1953.

9. EP: Meeting of Copperbelt urban court members, 5 April 1950.

10. EP: Case 5, Lupashi vs. Nyachinyama, Ndola, 1950.

11. EP: Second urban court members' conference, 13–16 April 1953.

12. EP: Discussions with court assessors, 1950.

13. ZA/Sec 2/406, vol. 3, 1943 (Cited in Chanock 1985, p. 208).

14. Commissioner of Police, Annual Report, 1954 (Lusaka, 1954).

15. Roan Antelope (RA) file 7: Town officer to acting personnel manager, 24 January 1964.

16. EP: Case 6, Maliria vs. John Konde, Ndola, 1950.

17. EP: Mfula vs. Simfukwe, Ndola, 1950.

18. EP: Dare Phiri vs. Belita, Chingola Urban Native Court, 1 February 1951.

19. EP: Discussion with court assessors, Ndola, 6/10/50.

20. EP: Case 32, Ndola, 1950.

21. Mikomfwa is a government township in Luanshya.

22. EP: Case 22, Ngumbo vs. Ngumbo, Mufulira.

23. EP: Case 5, Lupashi vs. Nyachinyama, Ndola, 1950.

24. EP: Case 37, Ndola, 1950.

25. EP: Case 1, (Urban Native (African) Court) UNC/2.

26. This happened in case 37.

27. EP: Case 17, Ndola, 1950.

28. EP: UNC/2, domestic relations, 4/3/54.

29. While polygynous households were rare on the Copperbelt (Epstein found one such household in his Ndola sample), divorce cases often blamed multiple wives for problems (Epstein 1981, pp. 36, 345–346). Wilson found 5.7 percent of 3500 women at Broken Hill were polygynously married (Wilson 1940, p. 64).

30. EP: Case 5, Lupashi vs. Nyachinyama, Ndola, 1950.

31. EP: Case 37, UNC/3, 4/12/53.

32. EP: Disucssion, Urban African Court, 21/5/54.

33. Same as note 30, above.

34. EP: Case 67, Ndola, 1950.

35. EP: Divorce, appeals court, Roberts vs. wife, 16/2/54.

36. EP: Divorce, appeals court, UNC/2, 5/11/53.

37. EP: Divorce, appeals court, Jaston Kalelemba vs. Bwalya Chalenga, 18/2/54.

38. EP: Case 37, Ndola, 1950.

39. EP: Case 50, UNC, 18/3/54.

40. EP: Case 17, UNC, 4/9/53.

41. EP: Case 69, Ndola, 1950.

42. EP: Case 78, Ndola, 1950.

43. EP: Divorce, appeals court, Mufulira, 1948.

44. EP: UAC/Court members, position of, on divorce, 21/5/54.

45. EP: Case 32, Ndola, 1950.

46. EP: Case 24, UNC/3, 22/9/53.

47. EP: Discussion with court assessors.

48. However, ethnicity and inheritance patterns affected divorce rates. Matrilineal peoples had significantly higher divorce rates than patrilineal peoples. For example, Mitchell's large Copperbelt survey reveals age standardized divorce rates of 31.44 percent for men and 16.91 percent for women among the Bemba. In contrast, the patrilineal Mambwe experienced rates of 5.3 percent for men and 11.8 percent for women (Mitchell 1986, personal communication).

49. EP: TRA/MIN, Discussion with trade union leader, Robinson Puta, 8/2/54.

50. EP: KAB/L/887, Ndola Family Budget Surveys, Hut 887, 8/3/56.

51. EP: Ndola, assault case, Mutale vs. Hasting Mutale; conversations with court members; assaults between women, usually over men, were common as well.

52. EP: Bilton Tonga vs. George Bunwe, Ndola, 1950.

53. EP: Court records from Ndola, Broken Hill and Mufulira, 1950-51.

54. Women frequently received private gifts from their lovers as well as possible cuts from compensation fees. EP: Case 3, UNC/3, 14/8/53.

55. EP: Case 40, UNC/3. 8/1/54.

56. EP: Case 602, 939, 906, and others, Ndola, 1951. Out of 37 adultery cases in 5 cases the wife alleged marital brutality drove her to adultery. EP: Adultery case records, Ndola, 1950.

57. EP: Case 403, Ndola, 1950.

58. EP: Case 415, Ndola, 1950.

59. EP: Case 72, Ndola, 1950.

60. EP: Matrimonial case 493, Kawiza vs. Lilai Mulamba, Kitwe UNC, 5/10/50; Case 32, UNC/3, 12/11/53.

61. For example, a husband sued a man for adultery because he shone a flashlight on the wife at the end of a welfare center movie. EP: Case 28, UNC/3, 26/9/53.

References

Ault, James. (1983) "Making 'Modern' Marriage 'Traditional': State Power and the Regulation of Marriage in Colonial Zambia." *Theory and Society* 12: 181-210.

Barnes, James. (1951) *Marriage in a Changing Society.* The Rhodes-Livingstone Papers, no. 20. London: Oxford University Press.

Brelsford, V. (1947) *Copperbelt Markets.* Lusaka: Government Printer.

Chanock, Martin. (1985) *Law, Custom and Social Order: The Colonial Experience in Malawi and Zambia.* Cambridge: Cambridge University Press.

Chauncey, G. (1981) "The Locus of Reproduction: Women's labour in the Zambian Copperbelt, 1927-1953." *Journal of Southern African Studies* 7, 2: 135-164.

Colson, Elizabeth. (1958) *Marriage and the Family among the Plateau Tonga of Northern Rhodesia.* Manchester: Manchester University Press.

Daniel, P. (1979) *Africanisation, Nationalisation and Inequality: Mining Labour and the Copperbelt in Zambian Development.* Cambridge: Cambridge University Press.

Dundas, C. (1921) "Native Laws of Some Bantu Tribes of East Africa." *Journal of the Royal Anthropological Institute* 51.

Epstein, A.L. (1951) "Some Aspects of the Conflict of Law and Urban Courts in Northern Rhodesia." *The Rhodes-Livingstone Journal* 12: 28-41.

———. (1953a) *The Administration of Justice and the Urban African.* Colonial Research Studies, no. 7. London: H.M.S.O.

———. (1953b) "The Role of African Courts in Urban Communities of the Northern Rhodesia Copperbelt." *The Rhodes-Livingstone Journal* 13: 1–18.

———. (1954) "Divorce Law and the Stability of Marriage among the Lunda of Kazembe." *The Rhodes-Livingstone Journal* 14: 1–20.

———. (1959) *Politics in an Urban African Community.* Manchester: Manchester University Press.

———. (1981) *Urbanization and Kinship. The Domestic Domain on the Copperbelt of Zambia, 1950–56.* New York: Academic Press.

Gluckman, Max. (1950) "Kinship and Marriage among the Lozi of Northern Rhodesia and the Zulu of Natal," in A.R. Radcliffe-Brown and Daryll Forde, eds. *African Systems of Kinship and Marriage.* London: Oxford University Press.

Gouldesbury, C. and H. Sheane, (1911) *The Great Plateau of Northern Rhodesia.* London: Edward Arnold.

Harries-Jones, P. (1964) "Marital Disputes and the Process of Conciliation in a Copperbelt Town." *The Rhodes-Livingstone Journal* 35: 29–72.

———. (1975) *Freedom and Labour: Mobilization and Political Control on the Zambian Copperbelt.* Oxford: Blackwell.

Keller, Bonnie. (1978) "Marriage and Medicine: Women's Search for love and luck," *African Social Research* 26.

Mitchell, Clyde. (1957) "Aspects of African Marriage on the Copperbelt of Northern Rhodesia." *Rhodes-Livingstone Journal* xxii: 1–30.

———. (1961) "African Marriage in a Changing World." In *Marriage and the Family: Report of the Annual Conference of the Northern Rhodesia Council of Social Services.* Lusaka: Government Printer. 1–14.

———. (1963) "Marriage Stability and Social Structure in Bantu Africa." *International Population Conference Proceedings, New York.* Tome II. London: 255–262.

Parpart, J.L. (1986a). "Class and Gender on the Copperbelt: Women in Northern Rhodesian Copper Mining Communities, 1926–1946," in C. Robertson and I. Berger, eds. in *Women and Class in Africa.* New York: Africana.

———. (1986b) "The Household and the Mineshaft: Gender and Class Struggles on the Zambian Copperbelt, 1924–66." *Journal of Southern African Studies* 13, 1: 36–56.

Passmore. (1956) "Report on the Loafer Problem on the Copperbelt." University of Zambia, Institute for African Studies.

Perrings, Charles. (1979) *Black Mineworkers in Central Africa.* New York: Africana.

Poewe, K.O. (1978) "Matriliny in the Throes of Change: Kinship, Descent and Marriage in Luapula, Zambia." *Africa* 48, 3 and 4: 215–218 and 353–367.

Powdermaker, H. 1962. *Copper Town: Changing Africa.* New York: Harper and Row.

Republic of Zambia, (1973). *Census of Population and Housing 1969, Final Report.* Volume I, Total Zambia (Lusaka: Central Statistical Office).

Richards, Audrey. (1939) *Land, Labour and Diet in Northern Rhodesia.* London: Oxford University Press.

———. (1940) *Bemba Marriage and Present Economic Conditions.* The Rhodes-Livingstone Papers, no. 4. Livingstone: The Rhodes-Livingstone Institute.

_____ . (1956) *Chisungu: A Girl's Initiation Ceremony Among the Bemba of Northern Rhodesia*. London: Faber and Faber Ltd.

Saffery, Lynn. (1943) *A Report on Some Aspects of African Living Conditions on the Copper Belt of Northern Rhodesia*. Lusaka: Government Printer.

Schuster, I. (1979) *New Women of Lusaka*. Palo Alto: Mayfield Publishing Company.

Staudt, K. (1986) "Feminist Anthropology and the Construction of Gender in Colonial Africa," mimeo.

Wilson, Godfrey. (1942) *An Essay on the Economics of Detribalization in Northern Rhodesia*. Part II. The Rhodes-Livingstone Papers, no. 6. Manchester: Manchester University Press.

Wilson, Monica. (1977) *For Men and Elders: Change in the Relations of Generations and of Men and Women among the Nyakusa-Ngonde People 1875-1971*. New York: Africana.

Wright, Marcia. (1982) "Justice, Women, and the Social Order in Abercorn, Northeastern Rhodesia, 1897-1930," in M.J. Hay and M. Wright, eds. *African Women and the Law: Historical Perspectives*. Boston: Boston University , Papers on Africa, VIII.

7

Domestic Labor in a Colonial City: Prostitution in Nairobi, 1900-1952[1]

Luise White

For reasons I myself do not fully understand, Africanists and feminists do not always read each others' more theoretical work. In practice this has meant that historians of migrant labor frequently ignore the fact that the support work of capitalism—the reproduction of labor power—takes place in the most intimate of relationships, while participants in the housework debate often seem unaware that in many places the generational reproduction of labor power—family life—takes place hundreds of miles from the worksite. In African studies, this has left us with no serious analyses of how urban labor subsisted day in and day out: some scholars have left us with the impression that those laborers fortunate enough to have jobs returned to them, day after day, by stamina alone (van Zwanenberg 1972, pp. 165-203; Stichter 1975-76, pp. 45-50). Urban wage labor was described almost in a vacuum: it was related to the depredations of international capital but not, somehow, to regular meals.

Prostitution in Nairobi was and is domestic labor. As such it only exists in relationship to wage labor. It is the work that replenishes—supports, flatters, feeds, idealizes, entertains, appreciates, reinforces, and relaxes, in some combination—men who work. Prostitution replenishes occasionally that labor power which is expended daily in pursuit of a wage. In other words, prostitutes in Nairobi (or anywhere else) do not provide these services regularly to men who cannot pay for them. Nairobi prostitutes sold domestic tasks individually or as a set; sexual intercourse may have been the most frequently demanded task (although this may not have been true for the late 1920s), but it was certainly not the only domestic task sold singly. Prostitutes routinely exchanged hot meals, cold snacks, bathwater, beverages, and an ideology of sexual relations in which women were unfailingly deferential and polite for a portion of a man's wage.

While scholars and reformers may not have understood, African wage laborers and colonial administrators knew exactly what prostitutes did: they renewed a labor force while depressing wages and saving municipalities thousands of pounds on African housing. In 1909, Norman Leys, then of the Kenya Medical Department, wrote that prostitution consisted of "home comforts," needed by Africans as much as Europeans.[2] In 1938 Nairobi's Municipal Native Affairs Officer was more exact:

> 25,886 males employed and living in Nairobi have only 3,356 female dependents in the town. This is a proportion of just over 1 to 8. A demand arises at once for a large number of native prostitutes. . . . The immigration into Nairobi of young Kikuyu girls is continually mentioned by the Kikuyu Native Councils urging that steps be taken to stop it. The position there is again aggravated by the lack of proper native housing; whereas the needs of eight men may be served by the provision of two rooms for the men and one for the prostitute, were housing provided for these natives and their families, six rooms would probably be needed.[3]

Prostitution in Nairobi, then, existed in close relationship to two seemingly unrelated categories: male laborers and housing. Indeed, my own data suggest that sex ratios have very little to do with prostitution and that supply and demand alone do not determine prostitutes' earnings. It was not simply the unbalanced urban sex ratios that resulted in prostitution, but rather the low-wage migrant labor system, in which men did not earn enough to purchase long-term domestic services. They could, however, afford short-term ones. Given this context of high demand for prostitutes' services, the level of their earnings was always high relative to unskilled wages. But earnings did fluctuate over time according to economic conditions: according to the number of men employed in town, the level of their real wages, and the rents they paid.

All Nairobi prostitutes were self-employed; there is no evidence, oral or written, of anything that could be construed as pimping. Men and women's access to housing, rather than their relative numbers, determined how prostitution was conducted and made profitable. Because the colonial city was segregated, and because African housing, legal or illegal, was insecure, Nairobi prostitutes developed labor forms that expressed a woman's relationship to her place of residence. Why one prostitute called out to men in the street while another waited at home for more discreet clientele had nothing to do with a woman's personality; it had to do with her choice of a labor form for her prostitution and her access to housing. These factors determined her behavior, not her personal insecurities.[4]

There were three forms of prostitution in colonial Nairobi; two, at least, emerged out of specific crises in East African peasant production. The oldest is the *watembezi* form (from the Swahili verb, *kutemba*, "to walk") in which the sale of domestic labor takes place somewhere other than the woman's place of residence. This form was at least as old as Nairobi (which was founded in 1899), when it was begun by women whose family fortunes were ruined in the rinderpest epidemics of the 1890s. Obviously, this was the only form of prostitution in which homeless women could engage, but many women earned their first month's rent through watembezi prostitution and continued to practice the form, which was characterized by high risks and high profits.

Malaya is the Coastal, proper Swahili word for prostitute and seems to have been first used in contradistinction to watembezi to describe the form that was developed in the New Native Location of Pumwani shortly after it opened, in about 1923. The malaya form mimicked marriage and so conformed to the by-laws of early Pumwani. In the malaya form all sales of domestic labor were made inside the woman's residence. The form was thus predicated on the fact that women had rooms to which men might also purchase access: in Pumwani, it was the men who walked the streets looking for women. Malaya prostitutes were generally women who came from impoverished farming families. They tended to earn more than unskilled male laborers, over longer period of time.

The *Wazi-Wazi* form is so called becaue it has been historically associated with Haya women from the Bukoba District in Tanzania, who were called Waziba in Pumwani because that is what the Ganda already working there called them. My evidence suggests that the form was started by Ganda and Haya women in the early 1930s who were attempting to restore their families' wealth in the face of plummeting cash crop prices. Like malaya prostitutes, Wazi-Wazi women had rooms, but they initiated the sale of domestic labor while sitting outside their rooms.

Obviously a woman could practice all three forms in the course of her working life, and many women practiced two forms in their careers. This does not mean that the forms were stages leading to one another, so that there was no hierarchy of earnings or status or respectability based on the forms of prostitution. At different times in Nairobi's history, one form or another dominated local prostitution, but that was because of the level of male wages and the cost of urban housing, not anything intrinsic to the forms themselves.

Watembezi Prostitution, 1900–1924

The ecological disasters that ushered in the twentieth century in what is today Kenya drove thousands of young men and women to towns in search of food, protection, and work. The watembezi form, clearly practiced by men and women at the turn of the century, (Jackson 1930, pp. 325–

26) was one way this mobility became a means of capital accumulation. As early as 1899, explorers complained that the Masai were dying out because of venereal disease and that coin had driven out trade goods in the towns already dominated by Masai "loose women."[5]

There is no question that the watembezi form dominated Nairobi prostitution in the years before Pumwani was built; my evidence indicates that women generally combined the watembezi form with trade or farming, "traditional" women's work. One woman, who was to become a propertied malaya woman, spoke of working on her family's farm in Kileleshwa and meeting European men there: "it was extra money, we went to pick beans and had a man in secret. Sometimes a woman would go there just for the men, she would take a *gunnia* so that no one would be suspicous, it looked like she was going to pick beans, and she would use the gunnia as a blanket. . . ."[6] Other women, in all likelihood those without a family presence and clearcut land rights in Nairobi, cut wood in the forest and sold it and sexual relations in the African villages and European suburbs that checkered the nascent city as early as 1905.

It was during World War I that professional prostitutes emerged for the first time in Nairobi, when watembezi women rented space behind stalls in the Indian Bazaar to which they took soldiers during the day. These women lived off earnings that came entirely from prostitution.

Such professional watembezi prostitution continued for many women after World War I, largely because these women were denied access to cultivation. In 1921 and '22 the only legal site in which Nairobi Africans could live, Pumwani, was built on what was later admitted to be the poorest soil in the city.[7] As a result, Africans' legal access to adjacent arable land was effectively curtailed. The wealthier African village of Pangani was scheduled for demolition in 1919, but was not fully evacuated until 1939. Watembezi prostitution persisted outside Pumwani in the early 1920s; it was almost invariably recorded in written evidence and thus it is difficult to ascertain the exact nature of these women's conduct.[8] Their appearance on Nairobi's streets, however, was short-lived because their occupation of areas designated for Indian residential use coincided with the city's ability to finally create "desireable residential areas" out of the habitations of African entrepreneurs. These women were literally harassed off the newly paved and improved streets of the colonial city for a few years as the Medical Department took action against "insanitary premises."[9]

Malaya Prostitution in Pumwani, 1922–1939

Malaya prostitution emerged in Pumwani by 1923 at the latest. This form offered the most extensive set of domestic services for sale—clients essentially leased whatever was in the woman's room: food, utensils with which to cook and eat it, bedding, bathwater, and the companionship and body of the woman herself. Men seemed to have quickly learned to rely on malaya women not only to provide an illusion of domestic life, but a safe place for it to take place. A woman who came to Pumwani in about 1925 said that men came to her first residence because "they knew it was a safe place because the owner had no husband to beat them."[10]

Malaya women from the early 1920s insisted that they did not ask to be paid until the man was about to leave.[11] This enabled them to charge for all the services and labor they had performed, although it sometimes meant that a woman would not be paid. Malaya women stated a preference for night long visits, which were quite profitable when women could charge for all the food and services provided. Many women described their expertise at this in terms of religion: they, and Pumwani's male propertied elite, were Muslim, and this, several malaya women suggested, made their domestic skills better than those of other women.

> There was a very special way Muslim women were supposed to behave
> . . . she must bring water and wash the man first, then herself, and she will
> cook, food if it's morning, tea in the afternoon, but it is not right for Muslims
> to go from sex to cooking, they must wash before they touch anything.
> . . . In those days it depended on the way you talked to a man. If you are
> hungry and don't tell the man how hungry you are, he can't understand.
> . . .[12]

Starting in the late 1920s, however, when Nairobi entered a construction boom equal to that of the late-1940s, malaya women became available for brief daytime visits—for tea, for a chat, or for sexual relations (which do not seem to have been in any greater demand than tea), in exchange for about ten to thirty-five percent of what they might expect to receive from a man who spent the night with them. The income from what became known as "short time" in Nairobi was not derived from specific values for specific services; it was cajoled out of their customers:

> If you spoke to these men, and told them about yourself, and kept your
> house clean, and gave them bathwater after sex, he would give you a few
> more pennies, and if he liked you he would come again, even to greet you,
> and you would give him tea, and then he would have to give you maybe 75
> cents. . . . If a man knew you and came to you regularly, he would give you
> a shilling, but if he was a stranger to you it would be 25 cents.[13]

The economics of the daytime visits are not difficult to understand. By 1929, the Native Affairs Department—not a body noted for letting the unemployed African go unnoticed—reported that fully 80 percent of the city's 25,000 Africans were in paid employment (Hake 1977, p. 44). Most of these men earned around Shs. 12/per month and a few skilled workers earned as much as Shs. 30/; rents in Pumwani and Pangani were between Shs. 2/50 and Shs. 5/ and many men shared rooms, thus decreasing their individual expenditure on rent and increasing their real wages. Moreover, starting in 1928 commodity prices began to fall: bread and flour decreased 12 and 25 percent, respectively, while sugar dropped to two-thirds of its 1924 price; only milk increased in price, costing just over 25 percent more than it had in 1924 (Cowen and Newman 1975, pp. 6–7). On balance it was very economical for malaya women to provide tea with milk and sugar, and bread, for the men who visited them in the daytime—men who were purchasing not only the snack, but its service. An ideology of gender relations emerged in which men earned and women provided.

In the late 1920s, men paid between Shs. 3/-Shs. 3/50 to spend the night with a malaya woman; generally the price was higher when the woman provided breakfast.[14] While monthly incomes were hard to estimate— Nairobi's labor requirements were seasonal—it seems likely that in the dry seasons when employment was highest, 1920s malaya women would have earned at least Shs. 5/50 per week, or Shs. 22/ per month at a conservative estimate. This was well in excess of unskilled workers' wages. Indeed, by the 1930s several successful malaya women began to purchase urban property, which years later they were to pass onto their own children, or if childless, to young women they had befriended while in town. They made legal efforts to guarantee that their property would not be transmitted back to the patrilineages into which they were born (White 1983, p. 178). These women established themselves as independent heads of households that sometimes became lineages. "At home what could I do? Grow crops for my husband or my father. In Nairobi I could earn my money, for myself."[15]

In the 1930s, Nairobi's labor requirement dropped sharply, and Nairobi's population declined from 32,000 in 1929 to 28,000 in 1930 and to as low as 23,000 in 1933. It rose slowly back to 28,000 in 1936 and then skyrocketed to 38,000 in 1937.[16] Although wages for skilled workers in particular declined, rents increased to about Shs. 7/-Shs. 8/ by 1934, and the number of potential tenants decreased, so that the workers who remained in Nairobi paid an increased proportion of their earnings on rent. Fewer men with less spending money brought a retrenchment in the domestic services malaya women provided. By 1933 at the latest, malaya women began to ask for cash in advance—sometimes losing money as they provided breakfast and bathwater to the men who spent the night—because "I had no reason to trust the man."[17] Commodity prices decreased during the depression as African food production increased, and men may have been less willing to pay prostitutes extra for food they could otherwise buy for less, or bring from home for free. Some malaya women from the 1930s said they cooked food for a man only if he brought it as a gift, something men frequently did to lower the price for a night-long visit. Men from the 1930s included foodstuffs in the price they paid, which was frequently less than the amount paid by men who did not bring food.[18] Nevertheless, there was in the early 1930s a general belief among young women in Nairobi that prostitution was a more reliable source of income than marriage. "In those days," said a woman born in Nairobi in about 1912 who first practiced the malaya form in 1930, "it was not dangerous to be a prostitute, any woman could be a prostitute. That was our job, to go with men to get money."[19] By the mid-1930s, short-term sexual relations paid for in advance became commonplace in Pumwani and were perfected in a part of Pumwani that was to become known as *Danguroni* (a local pronunciation of the Portuguese work for brothel), where women encouraged a high volume of anonymous encounters in which the men would be safe and the women well compensated.

> In those days . . . men used to fear that you were the wife of somebody and they would be beaten up if they entered your room, so if they talked to them through the door, and he trusted you . . . then you could let him in and you would have sex. After that the man would call a friend and tell him 'now I can tell you where to find a woman' and . . . if that first man . . . had to pay me Shs. 5/ for short-time, then all the friends of his I had sex with that same night would have to pay me Shs. 5/.[20]

Such prostitution became especially profitable as the number of men working in Nairobi increased beyond the capacity, and the costs, of African housing. Rents increased faster than wages in the late 1930s, and African workers could only minimize the cost of living by grossly overcrowding their rooms. The men who provided unskilled labor for construction projects often slept on the building sites themselves, and between 1933 and 1938 12,000 men were arrested for living outside Pumwani (Stren 1978, p. 191). For such men, unschooled in the Muslim mores of Pumwani, Danguroni offered a safe place to find women; the money they saved on rent probably enabled them to purchase the foodstuffs that a more genteel form of prostitution, available elsewhere in Pumwani, would have provided.

Watembezi Prostitution, 1928–1938

In 1928 there were two distinct kinds of watembezi activities in Nairobi: one took place in Pangani and Pumwani, while the other was practiced on those thoroughfares of Nairobi where it was illegal for an African to walk after 7:00 PM. The watembezi form practiced in Pumwani and Pangani was reminiscent of that practiced in Nairobi prior to World War I, a prostitution that was tacked onto the sale of agriculture produce. Late in 1928, the chief in the Kikuyu area close to Nairobi requested that the Local Native Council take action against those women who were going to Nairobi ostensibly to sell vegetables, but who in fact were prostituting themselves there, sometimes daily.[21] One woman who lived in Pangani confirmed this:

> . . . long before Pangani was torn down, there were women who came . . . and went with men there and then at night they would walk back home to Kiambu or Limuru, and only those of us who saw them in town knew what they really did. . . . Some of them did bring vegetables to sell, and they never stayed overnight in Nairobi.[22]

I have no evidence of such watembezi activity after 1930, but given the increasing land pressures in Kiambu and the number of Kikuyu who were for the first time without access to their relatives' land in the late '20s, many women would have had to replace agricultural incomes with illicit ones.

The watembezi form that dominated the streets leading out of central Nairobi was characterized by short-term sexual relations at high profits for the prostitute. Young women tended to come directly to River Road and Racecourse Road from the rural areas; they knew, or soon learned, that it was illegal for Africans to rent rooms there and paid extortionate rents for the privilege, about Shs. 50/ for a single room. Women could only afford such accommodation if they shared rooms, which meant that at least part of the time, watembezi had to be out of the streets both to find customers and to give their roommates privacy. This of course made these women exceptionally vulnerable to arrest or police harrassment but at the same time it gave them daily access to Europeans, Indians, and the Africans who worked in the town. Given the high proportion of Africans in wage labor in 1928, and the density of a motoring white male population (Hake 1977, p. 27), the watembezi form in River Road and Racecourse Road paid off very well, and rapidly, for these women who survived the risks.

The automobile provided a new dimension of solicitation—the driver could approach the woman—so that women did not have to call out to men and thus risk arrest. Many watembezi told police that they had just come down to the road to look at the cars.[23] Men in cars, moreover, could provide some aspects of domestic service, such as a place for sexual relations to take place, the cost of which would not be reckoned as part of the woman's payment. "River Road," said a woman who began watembezi prostitution there in the late '20s, "was a place for finding men with cars." She also maintained that it allowed her a wider variety of salaried clientel than she could find anywhere else in Nairobi:

. . . you could only have a regular Asian [Indian] boyfriend in River Road. In Pumwani the women stayed like wives, they went with whoever came to them. In River Road you could walk back and forth, back and forth, from Government Road to Kariokor, and you could find all kinds of men that way, white men, Asian men, African men with jobs in the Municipality and the Railroads, they all had to pass that way, and you could call to them and take them right to your house because your house was so near, women in Pumwani didn't see white men or Asians just strolling by.[24]

The depression only slightly reduced the profitability of the watembezi form in River Road. Indeed, by 1933 or '34 the woman quoted above had sufficiently overcome her objections to Pumwani to build a house there with her earnings. Prostitution and rents in the River Road area increased as the number of laborers did in Nairobi in the late 1930s.

The recovery of the late 1930s involved some small demand for female wage labor, especially in domestic service, particularly childcare. The wages for women were commensurate with those paid men in domestic service. Most servants were housed on their employers' premises. One Kikuyu woman and her friend left agricultural wage labor in Murang'a

. . . to go to Nairobi because my friend said that she had gone with European men and they paid a lot of money, and the best place to find money would be Nairobi. But when we came here, she got a job as an *ayah* for Asian people and they gave her a room . . . so we both lived there and every morning she would wake up early and go to work, while I would go out and look for men.[25]

Since prostitution generally paid better than domestic service, many women went from the watembezi form to childcare back to watembezi prostitution, while still others supplemented their incomes with streetwalking. Women with jobs, especially those with part-time jobs, were able to walk the streets with impunity, since they carried the documentation that would prevent arrests. Many such women abandoned the watembezi form when they lost their jobs, and moved indoors to the relative safety of the malaya form in Pumwani.[26]

Wazi-Wazi Prostitution in Pumwani, 1934–1946

Worldwide, the Wazi-Wazi form seems to develop at specific points in the economics of migrant labor, when there are thousands of men employed primarily in unskilled labor who are primarily immigrants. In Nairobi, however, the form was and is practiced in Pumwani alone (White 1985). There it allowed women a degree of visibility denied practitioners of the malaya form, without removing them from the relative security of legal

African housing. In this form, women sat outside their rooms, or on the verandah of the house in which they had rooms. As in the malaya form, women had rooms to which men could also purchase access, but all similarities ended there: Wazi-Wazi women lived in mean circumstances, even by Pumwani standards, outside rooms (built onto the original mud and pole structure, especially in the late 1930s) measuring often four by eight feet, or they shared larger eight feet by eight feet rooms. They rarely provided any food at all. Moreover, Wazi-Wazi women called out their prices to the men who passed by, and this and their visibility scandalized the quieter Muslim practitioners of the malaya form. It also eroded the profits earned by malaya women in the years of high employment in the late '30s.

The Wazi-Wazi form is so called because of its historical association with Haya women from the Bukoba District in northwest Tanganyika; these people were called Ziba in Nairobi, after the Ganda name for them. The form was started by the daughters of recently impoverished and indebted cash crop producers in Tanganyika and Uganda (for whom the world price of coffee had dropped ninety percent in 1930), whose prostitution refinanced and subsidized a generation of fathers. They never acquired property in Nairobi for themselves, and eventually returned home to the respect normally accorded "the economic saviours of their families" (Sundkler 1945, p. 254; White 1983, pp. 186–190; see also Swantz 1985, pp. 73–77).

Obviously, the practitioners of such a labor form were not overly concerned with maintaining the decorum of the Pumwani community. The Wazi-Wazi women who established neighborhood price boundaries by calling out their prices to passing men scorned one another by "shouting 'you sell your vagina for Shs. 1/' or saying it was 50 cents, or even 25 cents." A Nairobi-born woman who was in her teens in the mid-1930s recalled

These Gisu and Waganda women used to sit in front of their doors and if they saw a man passing, they just called out to him, without any shame at all. And those Waziba women, from the time they started doing this work of prostitution . . . they used to sit in their doorways calling out to men . . . . At that time when a man went with a woman and tried to cheat her they would fight, and then the woman would start screaming for help. . . . The Waziba women . . . started doing shameful things when a man tried to cheat them, they would start to abuse the men, and they would ask their neighbors to come, and they would have a fight, in front of everyone. . . . At that time a woman would risk her blood to get her money.[27]

This urgency had a tremendous impact on the malaya women who were scandalized by it. They understood that this was family labor, the work of daughters who in the town behaved as sisters:

In those days it was the Waganda and Qaziba women who were rich and they didn't buy any plots here in Majengo. If one of those women died by mistake here in Nairobi her neighbors and friends who had come with her to Nairobi and lived in the same place with her would take her clothes and money and also her body and go straight to her home. Those Waziba women wouldn't have graves here in Nairobi, they'd go straight back to their own country. These women . . . were sending money home to their own countries every week. . . . When an Mziba woman died a neighbor would take the body and all her property to her father's house.[28]

But the main characteristic of Wazi-Wazi prostitution, the idea of brief sexual relations without any additional domestic services paid for in advance, the establishing in effect of a fixed value for sexual intercourse, had a strong impact on Pumwani prostitutes. A Tanganyikan woman (not a Haya) who first came to Pumwani in 1935 abandoned the watembezi form for Wazi-Wazi prostitution in about 1938, the year in which commodity prices began to rise at an average annual rate of seven percent (Cowen and Newman 1975, p. 7). She explained that she requested a low price in advance "Because that way I was sure to get a man, I was sure to eat that day, and I got the money in advance so no one could cheat me." By the late 1930s, then, the Wazi-Wazi form had become a source of security in Pumwani. She went on to contrast the Wazi-Wazi form with what she considered the much riskier old-style malaya form:

I thought that was a bad idea. There was one old woman [who] would wait until one of her boyfriends came for the whole night, then would go with him and feed him in the morning, she would even make bathwater for him, and you know water is not free in Pumwani, and then . . . she would get Shs. 2/ for the whole night, she would take whatever the man gave her, and in those days water and tea and food would cost 75 cents at least. So I ask you, which is better, to get Shs. 1/ for sure, and when I was lucky, when I was young and lucky, I could make Shs. 3/ or Shs. 4/ in one day, or to wait with food and milk for a man who might only promise me money?[29]

The overall effect of the Wazi-Wazi form on Pumwani prostitution in the 1930s was that the price for short daytime visits rose slightly and the price for night-long visits dropped slightly. Wazi-Wazi women attempted to minimize services and costs (bathwater, tea) by setting a price for sexual relations in advance. By setting that price between about Shs. 1/ to an ambitious Shs. 2/ throughout the 1930s, when unskilled laborers earned between Shs. 14/ (if privately employed and probably not housed at all) and Shs. 9/-12/ per month if they worked for the state (with housing), and when house servants' wages decreased by a full one third (Cowen and Newman 1975, p. 5), they set a price that was within the reach of most of the members of the workforce. The low price set in advance—indeed, often yelled out into the street—probably kept some Wazi-Wazi women from asking for less money.

The Wazi-Wazi form was exceptionally convenient for women with small children who had no nearby kin: they did not have to arrange for childcare (which they could be called on to reciprocate) if they were only going indoors for a few minutes. The amount of labor time involved in Wazi-Wazi prostitution had advantages to married women who practiced the form as well.[30] Starting in 1938, however, rents began to rise dramatically—to as much as Shs. 36/ per month—and commodity prices increased steadily. From 1939–45, the price of maize and potatoes rose at about 14 percent a year (Cowen and Newman 1975, p. 7). This made it very difficult for Pumwani prostitutes to profit by providing food. This in turn meant that prostitutes of all forms who had children or dependent spouses living with them need a larger and larger proportion of their profits to maintain the subsistence of their households. Whatever preferences individual women might have had, the full-scale domestic services of the malaya form became viable mainly for childless women.[31] But the women who demanded payment in advance did not always prosper. During World War II, when the factors that made the Wazi-Wazi form so profitable were intensified, Wazi-Wazi women continually underpriced themselves to the well-paid African soldiers and Italian prisoners-of-war who visited them in an area that became "like home to them."[32]

Prostitution During World War II, 1939–1945

World War II posed a dilemma for Nairobi's prostitutes. They were in the business of providing domestic services—sleeping space, companionship, sexual relations, and food—to working men without a family presence in the town, yet the tens of thousands of those men who during the war were available to prostitutes had their food, companionship, and housing taken care of by various branches of the armed forces. Kenya also housed about 20,000 Italian prisoners of war captured in Ethiopia and imprisoned in conditions undreamed of by the authors of the Geneva Convention. All of these men could be penalized for not taking advantage of military food and sleeping quarters, however much they might have wanted to purchase them for independent suppliers in Pumwani. The question is not only how did Nairobi prostitutes sell sexual relations to soldiers, but how did they sell the non-sexual tasks that accounted for their greatest profit. Nairobi prostitutes had to organize the sale of their labors in such a way as to sell domestic services to men who already had them provided. Moreover, they had to gain repeated access to those men for whom the state attempted to provide home comforts.

In the early years of the war, prostitutes did well, although not remarkably so, solely through the sale of sexual relations. In an era when Africans' incomes rose through increased production and black market trade (Anderson and Throup 1985, p. 338), juvenile watembezi prostitution dominated the thoroughfares around the Royal Air Force Base in Eastleigh, near Pumwani. African soldiers and POWs tended to converge on Pumwani in the late afternoon. Brawls frequently broke out, and by the end of 1942 Pumwani was declared off-limits to military personnel. For malaya and Wazi-Wazi women this meant, very simply, that the best paid men, making between Shs. 32/ and Shs. 45/ (Clayton and Savage 1974, p. 230ff.) in the city, could not come to them. One Wazi-Wazi women adjusted to this situation with remarkable ingenuity: she sold a specific form of domestic labor:

> The only way a European soldier could come visit you was to bring you some of his clothes and then a day like Saturday you would tie his clothes together and take them back to the camp, and you could go with him there, because the gate-keeper would let you in, because you would say that you brought the soldier his uniform that you'd repaired. Some women did this every week with their European boyfriends.[33]

But many other women lacked this kind of preferential access to army camps. They transformed a labor process to suit their needs, literally mobilizing the malaya form into a nightly watembezi prostitution at army and POW camps. They remained malaya prostitutes by day—and perhaps most importantly, in their own minds—but joined together in groups of between five and twenty to walk to the camps. By 1943 or '44 they were escorted part of the way there and back by their boyfriends, to protect them from thieves and presumably the police as well. Some of these women, particularly those who had been Wazi-Wazi prostitutes in Pumwani, tended to underprice themselves. European soldiers and members of the Central African King's African Rifles and West African Frontier Forces knew no Swahili, so bargaining was conducted by a show of fingers, and accounts of women who regularly received between Shs. 1/ and Shs. 5/ for sexual intercourse with foreign soldiers and POWs were commonplace, although one woman assured me that "a woman could make Shs. 100/ a week that way."[34] But the watembezi prostitution by malaya women showed how willing women were to update and mobilize key aspects of the form to increase and protect their profits to suit the special needs of a colonial city turned briefly into a garrison town (White 1986, pp. 255–273).

Nevertheless, during the war, the malaya form spread out from Pumwani. A few women, neither Muslim nor Nairobi-born, successfully exported the values of the peace-keeping, Pumwani malaya form to the outskirts of Nairobi. A watembezi women during the late 1930s, this woman moved to

Buruburu at the start of the war and built a lodging house and practiced a malaya form. She rented rooms there to prostitutes, and over a ten year period her net income from rent alone was Shs. 6000/. Her house was frequented by Italian POW truck drivers, and she described how she and her tenants organized the illusion of mongamous domesticity to minimize potential conflict and maximize potential profits:

> I myself was afraid of fights, so I only had one Italian boyfriend at a time, but some women used to have more than one . . . and such women would give each man a certain day of the week on which to come and on that day and that day only he would come. That's because the bedsheets those Italians stole for us had numbers on them, and the women with more than one Italian boyfriend would remove the sheets that didn't belong to the man and put on the ones that did on the day he was to come. This way there could be no jealousy, because when an Italian came he found the sheets with his number on the bed.[35]

During the war, many malaya women moved to Eastleigh and there practiced the watembezi form by day and malaya prostitution by night. Eastleigh was officially a Non-Native Area, legal housing for Somalis and Indians, in which Africans could live legally only if they were domestic servants whose employers could not afford to accommodate them in their own servants' quarters. The residence of Africans in Eastleigh, though technically illegal, did not come under official scrutiny until the wholesale arrests of the Mau Mau Emergency. This history of a continuous African presence in Eastleigh meant that African women could live there, albeit paying high rents, without arousing suspicion. These women began to practice full-service malaya prostitution for servicemen far from home, especially the Ghanaian soldiers who were often the subject of grim rumors of cannibalism and rape, but who a few women assured me were simply misunderstood:

> I used to cook for them . . . they ate beef, and they brought a lot of food for me in those days. . . . Most of the women who went with Golgos [Gold Coast, the colonial name for Ghana], they never went hungry, their houses were always filled with food. . . . People used to say Golgos ate each other, but this was not true, they fought with each other a lot. . . . If two ever met in my room, they would fight, and after that, the one who was there first, he'll cut off your ear, that is what they said they would do if they saw you with one of their friends. So when a woman went with a Golgos man, the most important thing she could do was remember completely the day of her appointment and not make an appointment with any other man for that day.[36]

In the quotations above, it was not the ethnicity or the quality of the men that determined anything at all; it was the organization of domestic labor, women's ability to provide a superficial appearance of monogamy and devotion that determined the ability of the malaya form—wherever it was practiced, whoever it was practied with—to provide profit during World War II.

Changes in Nairobi Prostitution, 1946–1952

The laboring population of Nairobi District had trebled between 1941 and 1945; these workers were historically most sensitive to increases in the cost of living, since they regularly purchased their foodstuffs (Stichter 1982, pp. 111–112). The special burdens and booms of the Second World War (Cooper 1981, pp. 33–39; Anderson and Throup 1985, pp. 327–345) ushered in an era of a steadily increasing cost of living, which resulted in simultaneous labor organization and job insecurity—a new norm of urban subsistence for workers in Nairobi's informal economy.[37]

In the late 1940s, the wages of African skilled workers were higher than ever before. Most African unskilled and semi-skilled workers earned between Shs. 20/-30/ a month (Cowen and Newman 1975, p. 13; Stichter 1982, p. 115), and there were also more of them than there had been before. Those in unskilled construction work, then booming, earned even less. Rents rose steadily, but not as steeply as they had done in the late 1930s: in 1946, the rents in Pumwani ranged from Shs. 13/-Shs.26/; in 1950 that range was from Shs. 16/-Shs. 32. Elsewhere, in the squatter settlements of Mathare, Buruburu, and Kamangware rents rose to between Shs. 20/-Shs/ 30/, largely due to the increased cost of building materials for even mud huts. Even in the houses owned illegally by African women in Eastleigh rents doubled and trebled to between Shs. 70/-Shs/ 80/ after 1947 because, as one former tenant reported matter-of-factly, the new owner "had to repay a loan."[38] Food prices increased as well.

The Nairobi Municipal Council sporadically built married workers' quarters in its housing estates in an attempt to stablize their labor force: in practice, whole families had to live on a single man's wage, in a single room or even only part of a room. Perhaps because of such conditions a number of married women with children to support entered the ranks of Nairobi prostitution in this period. Children took on a special importance during the post-war years. The ex-servicemen who visited prostitutes were, among other things, veterans of countless lectures on the evils of venereal diseases, including barrenness, and had a healthy respect for the role of sexual intercourse in transmitting disease.[39] Prior to World War II, it was generally held that men preferred to visit prostitutes without children because "Children make noise and want attention the first thing in the morning, and that is not what a man goes to a prostitute for."[40] But after the war "some men feared that a woman without children had gonorrhea and that they would get it, and these men would only go with a woman who had a child already."[41]

As we have already seen, no new forms were developed after the mid-1930s; instead, women combined the existing forms or practiced them in new places. In Eastleigh, full-service malaya prostitution was practiced at night by women who engaged in the watembezi form in the daytime in downtown, white Nairobi—sometimes taking a small child along with them. The women who did this went to great lengths to appear respectable:

> I went with men. I was jailed for not having a pass-book. . . . When I got out of jail this time I had to bribe a man Shs. 150/ to say he was my husband so I could get a pass-book . . . he was a cook for Europeans. I had known him for several years, not a permanent boyfriend but a friend, and I decided to ask him to do this for me because I knew he needed money, he had two wives at home.[42]

In Pumwani, women with small children did not practice the Wazi-Wazi form but changed the watembezi form to suit their needs and those of the laboring men who surrounded them in construction sites for MNC housing estates adjacent to Pumwani. By 1947/48, many women in Pumwani took to daytime watembezi prostitution, previously unheard of there, and solicited men at their workplace. They requested a cash sum that was payable in advance, and generally not subject to negotiation. This scandalized the older Muslim malaya women—"It was one thing to go in the alley to look for men. . . it was another to let your elders know you were doing this."[43] But one Christian woman who had come to Pumwani in 1945 justified this watembezi practice with great care and thoughtfulness:

> I used to go to Reuben's camp [site of Gorofani, next to Pumwani] to look for men. . . . I had a small child I always carried around, I couldn't afford to have an ayah, and a white man couldn't stomach the idea of you coming with a child to his house but the Africans were alright about it. . . . The price was Shs. 1/ for short-time and Shs. 5/ for the whole night . . . everyone knew the price. . . . Some women asked for more but others like me did not. It was like this, I didn't have any money and I needed to get some. If a man knew he could go with me for Shs. 1/, he would, and I wouldn't have to worry about bargaining for fighting or abuse, and if he knew that it was Shs. 5/ for the whole night he would pay me that in advance, and I wouldn't have to spend money on tea or anything in the hopes he would give me a few cents more in the morning. Don't you see, I had five children, I had lived with several men, and it was simpler for me to go with a man who knew in advance what he was going to pay.[44]

Such subsistence prostitution earned its practitioners sums that were well in excess of unskilled laborers' incomes. The watembezi prostitution of the woman quoted above earned her, by the most conservative estimate imagineable, Shs. 30/ per month; Shs. 45/ to Shs. 60/ were more likely. But for most of the years she visited construction sites she paid Shs. 23/ a month in rent, raised five children, and supported at least as many men through periods of unemployment and underemployment. She lived comfortably by Pumwani standards, and she was able to send her children through primary school; in 1977, she lived on her savings and contributions from two of the children who still shared her large room. But she was typical of the immigration prostitutes of the post-war era in that her relatively high income was not enough for her to invest in urban or rural property.

The daytime watembezi prostitution within Pumwani lowered the price for short-time throughout Pumwani. Between 1946 and 1952 the short-time price of Wazi-Wazi women ranged from Shs. 1/50 (the price for a pound of maize flour in 1950) to Shs. 3/ (the 1949 price of packaged milk).[45] The amounts were lower than those received by Wazi-Wazi women during the war years.

Many Wazi-Wazi women responded to their diminishing incomes by organizing other practitioners of the form to fix their prices by methods other than calling them into the street. This was in fact the first attempt by any prostitutes to establish and maintain a short-time price that they, and not their customers, had fixed. My earliest data on price-fixing comes from 1948; I suspect it might have begun a year or two earlier. It seems to have been initiated by Haya women but the actual price-fixing was organized through adjacent houses, not ethnicity. Wazi-Wazi women, including new practitioners from western Kenya and the housing estates nearby, talked to each other about how much they should be charging. A woman who came to Pumwani in 1948 described how the Haya women living in her house decided to raise their short-term prices, "We all agreed never to go for less than Shs. 2/. Most of the women who decided were Waziba, but we asked our neighbors to help us. Some of our neighbors were Luo women; some were Nandi."[46] They were also all Christian. Although my data suggest that in 1948 price-fixing was confined to a number of houses in west side of Pumwani, by 1950 it had spread to houses throughout Pumwani.[47]

The Wazi-Wazi short-time price influenced the prices paid practitioners of other forms in Pumwani in the post-war years. When a Muslim woman came to Pumwani in 1950 she practiced the malaya form; she was informed by the men who came that the short-time price was Shs. 2/.[48] My data suggest that the malaya women who had gone together to the military camps during World War II maintained their short-time prices at the levels established by Wazi-Wazi women after 1948. These women did not, however, regard their doing this as any kind of workers' solidarity; they saw it as sensible economics.[49] Nevertheless, the malaya women who had not gone to army or POW camps during the war generally did not observe the short-time minimum price, and in fact had fairly elaborate strategies about how to get more from men.

Conclusions

Historians generally regard African history (to say nothing of African labor history) and women's history as two discrete fields, related only by the research needs of some overzealous feminist scholars, not by the intrinstic nature of labor and women. But men and women have a great deal to do with each other, and the study of what happens when working men seek the companionship of equally hard working women is essential to an accurate picture of African urban history. The history of prostitution shows how working men lived and loved, how their urban subsistence was transformed into the reconstruction of rural lineages (in the Wazi-Wazi form) and independent urban class formation (by the malaya form). The labor forms of prostitution were determined by the availability of housing just as housing was planned as a mechanism for labor control but was often taken over by Africans themselves for purposes inimical to colonial notions of order (Cooper 1983, pp. 31–34).

In this essay I have shown how prostitution changed in relation to the changes in wage labor. When real incomes were high, women could sell a range of services: night-long visits, meals, bath water and companionship. Their profits were high. When real incomes were lower, such as during the depression of the early 1930's, demand for prostitutes' services was less, and the range of saleable services decreased. One indicator of the tight money situation was that women had to ask for cash in advance. Their profits decreased. During the depression, more short-term encounters became the most common form of prostitution, and after the depression, when demand increased in the late 1930's, both the short-term forms, watembezi and Wazi-Wazi, became quite profitable. They competed with the older malaya form, which became less profitable as prices of the commodities women needed to buy to provide domestic services increased. After 1946, business continued to flourish, although there was increasing competition, reflected in price-fixing efforts by some Wazi-Wazi women.

Prostitution is labor, work performed for, and paid for by labouring men: to situate the causes of prostitution in more abstract or larger categories simply obscures the concrete nature of prostitutes' work. The Nairobi prostitutes described here prove—as some feminist theorists have hypothesized (Molyneaux 1979, p. 5)—that the tasks we normally think of as constituting domestic labor can be individually sold to laboring men with very profitable results. My data suggest that there is nothing about prostitution in and of itself that interferes with profit; in fact, it is the proportion of a woman's income spent on childrearing that determined her rate of profit. For most of the colonial era, prostitutes earned almost twice what unskilled male laborers earned; as high commodity costs eroded larger and larger proportions of their incomes, the number of women who achieved the status of property owners declined, but the number who prospered, who made history, and good money as well, in the unsolicited and appalling conditions of the colonial experience did not.

Notes

1. Earlier versions of this essay were presented at the African Studies Association, Baltimore, Maryland, 1979, The African Studies Center at Boston University in April, 1980, and the Seminar in Atlantic History, Society and Culture at The Johns Hopkins University in November, 1984. I am grateful for the participants' comments, and especially to Edward Alpers, Frederick Cooper, Judith Geist, Margaret Jean Hay, John Lonsdale, and Richard Roberts for their comments, and to the editors of this volume.

2. Norman Leys, "Appendix to the Nakuru Annual Medical Rep. 1909: an account of venereal disease in the Naivasha District," Kenya National Archives (hereafter KNA): PC/NZA/2/3.

3. Eric St. A. Davies, "Some Conditions Arising from the Conditions of Housing and Employment of Natives in Nairobi," April, 1939, KNA: MD 40/1131, v. 1.

4. Most of the data that follows comes from my Ph.D. thesis, "A History of Prostitution in Nairobi, Kenya, c. 1900-1952," Cambridge University, 1986, and is based on seventy interviews conducted with former prostitutes and twenty-six of their clients. I am grateful to my assistants, Margaret Makuna and Paul Kakaire.

5. Professor Mackinder Diary, 20 July and 30 September 1899, Rhodes House, Oxford, RH MSS Afr r29.

6. Amina Hali, Pumwani, 4 August 1976.

7. A.R. Patterson, Dir. Med. Sanitary Services, to Chief Secretary, Nairobi, "Cost of Native Housing," 12 April 1940, KNA: MD 40/1131.

8. Letters from P.P.L.P. Nyanzi and another [signature illegible] to Patterson, Med. Officer of Health, "Re: Brothels in River Road," 17 July 1923' "Incest in Nairobi," 21 November 1923 and "Prostitution in Nairobi," 2 May 1924, and Census of Prostitutes in Nairobi, Con. Memo. 8/77/1, 11 October 1923 in KNA: MD 28/854, Venereal Disease in Kenya, v. 1.

9. Patterson to Chief Native Comm., 13 October 1923, KNA: MD 28/854, v. 1.

10. Kayaya Thababu, Pumwani, 7 January 1977. Housing owned by single women was simply that—housing owned by single women, and should not be thought an organized brothel; see White, 1983.

11. Kayaya Thababu; Himani bint Ramadhani, Pumwani, 4 June 1976; Miriam Wageithiga, Pumwani, 11 August 1976; Mama Abdullah Saidi, Pumwani, 2 March 1977.

12. Amina Hali.

13. Amina Hali.

14. Men's Price Sheets, compiled by Paul Kakaire and Luise White, Nairobi, February-April 1977.

15. Kayaya Thababu.

16. Nairobi Medical Officer of Health Annual Reports, loaned to me by R.M.A. van Zwanenberg.

17. Zeitun Ahmed, Pumwani, 14 June 1976.

18. Rehema binti Hasan, Pumwani, 21 December 1976; Zeitun Ahmed; Salima binti Osmoni, Pumwani, 17 June 1976; Men's Price Sheets.

19. Fatuma Makusidi, Pumwani, 20 May 1976.

20. Salima binti Osmoni, Pumwani. My data suggests that Shs. 5/ is an exaggeration, and that Shs. 2/-Shs. 3/ would be more accurate and a large sum by mid- and late-1930s standards.

21. LNC Minutes, 1925-31, 22-23 Nov. 1928, in KNA: KBU/LNC. I am grateful to Dr. Marshall Clough for giving me this reference.

22. Wanjiko Kamau, Mathare, 1 July 1976; see also Rehema binti Hasan, Pumwani, 21 December 1976.

23. Zeitun Ahmed; Wambui Murithi, Pumwani, 14 December 1976.

24. Wambui Murithi; see also Gathiro wa Chege, Mathare, 9 July 1976.

25. Wanjira Ng'ang'a, Mathare, 8 July 1976.

26. Wanjira Ng'ang'a; Elizabeth Nomsogei, Pumwani, 17 August 1976.

27. Margaret Githeka, Mathare, 2 March 1976.

28. Fatuma Makusidi, Pumwani, 20 May 1976.

29. Mary Salehi Nyazura, Pumwani, 13 January 1977.

30. Rehema binti Hasan, Pumwani, 21 December 1976; Asha Mohammed, Pumwani, 1 March 1977; Mary Salehi Nyazura.

31. Miriam Musale, Pumwani, 17 June 1976.

32. Miriam Musale.

33. Mary Salehi Nyazura.

34. Zeitun Ahmed, Pumwani, 14 Juen 1976; see also Rehema Monteo, Pumwani, 10 August 1976; Elizabeth Nomsogei, Pumwani, 17 August 1976; Sara Waigo, Mathare, 1 July 1976.

35. Wanjira Ng'ang'a, Mathare, 8 July 1976.

36. Tabitha Waweru, Mathare, 13 July 1976.

37. This may in fact be a characteristic of crises in casual labor; see Cooper, 1986.

38. Margaret Githeka; Mary Salehi Nyazura; Sara Waigo, Mathare, 1 July 1976; Beatrice Nyambura, Mathare, 8 July 1976; Naomi Kayengi, Pumwani, 3 August 1976; Rehema Monteo, Pumwani, 10 August 1976; Hawa binti Musa, Pumwani, 1 December 1976; Alisati binti Salim, Pumwani, 22 February 1977; Asha Mohammed, Pumwani, 1 March 1977; Hadija binti Nasolo, Pumwani, 3 and 8 March 1977; Tamima binti Saidi, Pumwani, 15 March 1977.

39. Kings' African Riles East African Cmmd., *Current Affairs*, 9 (1944) suggested this speech for commanding officers: "You have often blamed women for failing to bring forth children, and you have sent them back to their fathers. What should

be done to you, though, if you cause the woman to be sterile?" KNA: MD 28/ 854, Venereal Disease in Kenya, v. iv.

40. Hadija binti Karim, Pumwani, 16 February 1977; see also Kayaya Thababu; Fatuma Makusidi, Pumwani, 20 May 1976.

41. Lucia Kaseem, Pumwani, 28 March 1977.

42. Beatrice Nyambura; see also Margaret Githeka; Tabitha Waweru, Mathare, 13 July 1976.

43. Hawa binti Musa.

44. Eflas Negesa.

45. These figures were compiled from my interviews and notes graciously loaned to me by Michael Cowen.

46. Juliana Jacob, Pumwani, 17 March 1977.

47. Juliana Jacob; Elda Ayoo, Pumwani, 24 February 1977; Rebecca Okumu, Pumwani, 13 February 1977.

48. Tamima binti Saidi, Pumwani, 2 March 1977.

49. Asha Wanjiru, Pumwani, 23 December 1976; Sara Waigo; Mary Salehi Nyazura.

References

Anderson, David and D. Throup (1985) "Africans and Agricultural Production in Colonial Kenya: The Myth of the War as a Watershed," *Journal of African History.* 26: 327–345.

Clayton, Anthony and D.C. Savage (1974) *Government and Labour in Kenya, 1895–1963.* London: Frank Cass & Co.

Cooper, Frederick. (1981) "Africa and the World Economy," *African Studies Review,* XXIV, 2/3 (June/September): 1–86.

———. (1983) "Urban Space, Industrial Time, and Wage Labor in Africa," pp. 7–50 in Frederick Cooper, ed., *Struggle for the City: Migrant Labor, Capital, and the State in Urban Africa.* Beverly Hills and London: Sage.

———. (1986) *On the African Waterfront: The Transformation of Work in Mombasa, Kenya 1985–1963.* London: Yale University Press.

Cowen, Michael and J. Newman (1975) "Real Incomes in Central Kenya, 1924–70," mimeo.

Hake, Andrew. (1977) *African Metropolis: Nairobi's Self-Help City.* Brighton: Sussex University Press.

Jackson, [Sir] Frederick. (1930) *Early Days in East Africa.* London.

Mann, Kristin. (1985) *Marrying Well: Marriage, Status and Social Change among the Educated Elite in Colonial Lagos,* Cambridge: Cambridge University Press.

Molyneaux, Maxine. (1979) "Beyond the Domestic Labor Debate," *New Left Review.* 116 (July-August): 3–28.

Oppong, Christine, ed. (1983) *Female and Male in West Africa.* London: George Allen and Unwin.

Robertson, Claire C. (1984) *Sharing the Same Bowl: A Socioeconomic History of Women and Class in Accra, Ghana,* Bloomington: Indiana University Press.

Stitchter, Sharon. (1975–76) "Women and the Labor Force in Kenya, 1895–1964," *Rural Africana,* 29 (Winter): 45–68.

———. (1982) *Migrant Labour in Kenya: Capitalism and African Response 1895–1975*. Harlow, Essex: Longman.

Stren, Richard. (1978) *Housing the Urban Poor in Africa: Policy, Politics, and Bureaucracy in Mombasa*. Berkeley: Institute of International Studies, Research Ser. 34.

Sundkler, B.G.M. (1945) "Marriage Problems in the Church in Tanganyika," *International Review of Missions*, 24, 135: 253–266.

Swantz, Marja-Lisa. (1985) *Women in Development: A Creative Role Denied*. New York: St. Martin's Press.

White, Luise. (1983) "A Colonial State and an African Petty Bourgeoisie: Prostitution, Property, and Class Struggle in Nairobi, 1936–40," pp. 167–194 in Frederick Cooper, ed. *Struggle for the City: Migrant Labor, Capital and the State in Urban Africa*. Beverly Hills and London: Sage.

———. (1985) "Prostitutes, Reformers, and Historians." *Criminal Justice History*, IV.

———. (1986) "Prostitution, Identity, and Class Consciousness in Nairobi during World War II." *Signs: Journal of Women in Culture and Society*, 11,2: 255–273.

Van Zwanenberg, R.M.A. (1972) "History and Theory of Urban Poverty in Nairobi: The Problem of Slum Development," *Journal of East African Research and Development*. 2,2: 165–203.

8

Evading Male Control: Women in the Second Economy in Zaire

Janet MacGaffey

Although the formal authority structure of a society may declare that women are impotent and irrelevant, close attention to women's strategies and motives, to the sorts of choices made by women, to the relationships they establish, and to the ends they achieve indicates that even in situations of overt sex role asymmetry women have a good deal more power than conventional theorists have assumed (Rosaldo and Lamphere 1974, p. 9).

One of the issues addressed in this book is whether or not capitalist penetration and development in Africa has offered income generating opportunities for women to any significant degree. Writing on African women tends to stress the oppressive effects of Africa's peripheral capitalism (see Boserup 1970; Parpart 1987), but a brighter side to the picture exists. It receives little attention because urban women are as invisible in official reports and statistics as rural women. Many of the most successful and lucrative economic activities of women in towns and cities go unrecognized because they are carried out in the second economy, that sector of the total economy which evades the control of the state or in some way deprives it of revenue, and which is unrecorded and thus unmeasured in figures on the GDP or GNP.[1]

Women in Zaire suffer from severe political and economic disadvantages. They are underrepresented in the state, since very few hold politically effective office or leadership roles; lack of education hampers them in entering the professions and salaried employment; and regulations and the legal system reinforce male control of their activities and circumscribe their independence. Some women have, nevertheless, overcome the difficulties confronting them and have become wealthy and successful.

Income generating opportunities for women in Zaire have fluctuated in accordance with political and economic changes, from the colonial period through to the seventies and eighties. The penetration of capitalism under the colonial state, the political upheavals following independence, the indigenization of foreign capital, the weakening of the administrative capacity of the post-colonial state, and the recent rapid expansion of the second economy have all affected the economic position of women.

Women in Zaire made the most of the very limited opportunities available to them in town in the colonial period, so that although they found themselves in a very disadvantaged position as they moved into urban life, the role of independent autonomous woman existed even in this early period. In the political and economic troubles after independence, sheer economic exigency made women's earning capacity essential for the maintenance of a household, so that at this level many women again achieved the role of food provider and the degree of economic autonomy represented by their own incomes that they had earlier enjoyed in most rural societies.

In the economic system that has developed with the penetration of capitalism in Zaire, petty producers of food in rural areas and of goods and services in the "informal sector" of towns have reduced the costs of the labor force for capitalist enterprise, enabling the workers to survive on very low wages. The wage labor force is primarily male, while women have been largely confined to the "informal sector." However, this circumstance has provided opportunities for some women to achieve a degree of economic independence.

Following independence, the political troubles of the sixties and the economic crises of the seventies, caused by the Zairianization of foreign businesses, the fall in the price of copper and world recession, together with the predatory and parasitic activities of the post-independence dominant class resulted in the general breakdown of institutions. The consequence was a decline in the administrative capacity of the state and a weakening of its control over economic processes. This situation led to competition for new forms of access to resources, especially via the second economy. It weakened the legal and institutional basis of male control over women because the institutions of the state, which are also instruments of male control, by definition do not operate in the second economy. "As new resources enter a political field, both men and women seek advantage from expanded opportunities, and the course of change necessarily reflects the complex interplay of male and female tactics" (Collier 1974, p. 96). In the seventies and eighties, increasing economic opportunities for women meant that some of them achieved substantial capital accumulation; they expanded their businesses and invested in real estate, transport and plantations.

Colonial and Post-Colonial Oppression of Women

Belgian colonial rule and the penetration of capitalism changed the economic as well as the political base of the different ethnic groups in the

Belgian Congo. In this process women generally found themselves in a more disadvantaged position than formerly, as they did throughout most of Africa (Boserup 1970, pp. 53–57; see Henn this volume); their subordination formed an integral part of the new economic order.

The colonial economic system benefited from women's role as food producers because it enabled men to become wage earners without disrupting food production. By contributing to the support of male wage laborers and their dependents, women lowered the cost of this labor (Wilson 1982, p. 154; Boserup 1970, pp. 76–79). In Zaire, as in other colonial systems, women were excluded from the modern economy. The state allowed women only minimal education and restricted its content to domestic skills. Girls were given no instruction in French, the language necessary for economic advancement. Their post-primary vocational schooling was limited to teaching, home economics and domestic agriculture, rather than the commercial agriculture in which boys were trained. In 1948 a six year program of secondary education was started for boys, but only three years were offered to girls. In the mid-fifties girls were admitted to the six year program but by independence in 1960 only about 1.5 percent of them attended any post-primary school (Yates 1982). This "allocation of women and men to different levels and content of education is a mechanism which contributes to the process of sex segmentation and skills differentiation in the labor force itself" (Mbilinyi 1985, p. 179).

Women at independence thus found themselves at a disadvantage in the urban labor market. Since then the situation has changed little; it is still much more difficult for girls to complete secondary school or university than for boys (Schwartz 1966; T. Gould 1978). Women are much less likely than men to be fluent in the French necessary for professional employment. Very few women have wage earning jobs that would enable them to accumulate money from salaries, gain knowledge of business management or build up business connections by working in large enterprises. Furthermore, although the rights of women who do work are protected by the present labor code, in practice the law is seldom applied with regard to equal pay, family allowances and housing, and women are in fact discriminated against in all these respects (Manwana 1982, p. 81). A married woman who works does not get a housing or family allowance, even if she is separated or divorced or her husband unemployed (Wilson 1982, p. 163).

The subordination of women is strengthened by legal restrictions on their activities. A woman must have her husband's permission to open a bank account or to obtain a commercial license. Until 1962, a woman had to have the consent of her husband to sign a work contract. The new law says that a woman can engage her services *unless* her husband expresses opposition: if he writes to her employer and objects, her contract is voided (Manwana 1982, p. 77). Book III of the proposed new Civil Code on the family seeks to give a man control over his wife's wealth (W. MacGaffey

1982). Article 45 presumes the management of the wife's goods to be in the husband's hands, even though the marriage contract may specify separate ownership of goods by each spouse. A woman may only take over management in case of the proven incompetence of her husband (Article 520). She may manage goods acquired in the exercise of her profession, but her husband is allowed to take them over "in the interests of the household" (Article 492). Property resulting from the profits of petty trade can be seized by the husband's kin, if the husband gave his wife a small sum of money to begin her trading career. If there is no male adult child to inherit under customary law, a widow may be forcibly ejected from her home and lose rights to any property (Wilson 1982, p. 164).

Given women's lack of education and the legal constraints from which they suffer, the second economy offers more opportunities for them than the official economy. Many women began to achieve economic autonomy through its activities; some became extremely wealthy.

Women's Changing Role in the Urban Economy, 1945–1970

Despite the disadvantages facing women as they moved into urban areas, Jean LaFontaine found, in a study of Kinshasa (then Leopoldville) in 1962–63, that 14 out of the 46 parcelles (lots on which people could build) she investigated in the commune of Kinshasa were owned by women, indicating that already at this time the role of independent woman was an "established phenomenon of urban life" (LaFontaine 1970, p. 60). Both African men and women were able to purchase freehold rights to their plots after 1954, and many built additional rooms for renting to accommodate the city's rapidly expanding population. Plot owners thus had a valuable income source and some people were able to live entirely off their lodgers' rents. The fact that a woman could be the owner of a parcelle in her own right gave economic independence to some urban women and made it at least a possibility for others. Eight out of fifty-two unmarried, divorced or widowed women in the Kinshasa commune study lived entirely off income from their parcelles; six others lived off such income together with gifts from kin and proceeds from trade or prostitution (LaFontaine 1970, p. 60). Given the difficulties facing women the question is, how did they come to own these parcelles?

In 1945 Suzanne Comhaire-Sylvain found that women in Kinshasa earned money in two primary ways: through petty trade and through prostitution. Unlike West Africa, Zaire had no tradition of large-scale trade by women between rural and urban areas, but in 1945 about 300 women were engaged in commerce of food products, such as manioc, palm oil, drinks and fish. In a survey of 730 mothers of families, 20.41 percent carried on such commerce. Some travelled from town to town buying and selling wholesale;

others sold retail what they bought from wholesalers; others would walk half a day's journey into the countryside to buy food products which they would carry back to town to sell for a couple of days before returning to get more (Comhaire-Sylvain 1968, pp. 27–28).

When women first moved into towns and cities they found themselves worse off as their role came to be more narrowly defined in sexual and domestic terms than it had been in the rural society from which they migrated (Pons 1969, p. 219). Under colonial law they were only allowed to migrate to town as the dependents of a resident. They could not remain in town if their marriage ended, unless they had work. Few of them had any education in the early years and the only occupation that enabled them legitimately to remain in town was prostitution.

Luise White, in this volume, shows that women in Nairobi, denied access to cultivation, turned to prostitution. In Kisangani some women turned this situation to their advantage. Some earned money working as individual prostitutes, not for a pimp; others were the relatively independent mistresses of wealthy men, or "courtesans" with small changing sets of lovers (LaFontaine 1974, p. 99).[2] A study of Kisangani (then Stanleyville) in 1952 found a number of divorced or widowed women who had spent their young lives as prostitutes and had then become shopkeepers and landladies (Pons 1969, p. 248).

In Kinshasa these more independent urban women were already numerous by 1945. At that time, men outnumbered women seven to four; the majority of women were in the 20 to 40 years age group and had few children. Forty to fifty percent of the 8,000 women over 14 were unmarried; 1,000 of them were between 14 and 18 years, some engaged or living with their prospective husbands; about 100 were elderly and divorced or widowed; approximately 600 were prostitutes. This left about 6,000 women living in concubinage or temporary unions and listed as dependents of male partners or kinsmen. As concubines these women preserved their liberty; a lover was less able to be a despot and exercise control (Comhaire-Sylvain 1968, pp. 11, 23–25). Men did not like their wives working, mainly because they believed a wife's economic independence threatened their authority (Comhaire-Sylvain 1968, p. 32).

By the end of the colonial period, therefore, by making the most of the very restricted opportunities open to them, some women were already beginning to find means to achieve some autonomy. The new sources of income in real estate, petty trade and prostitution remained critically important to urban women during the turbulent politics of the sixties. The violent and anarchic situation following independence in 1960 gave rise to unemployment, rising prices, the decline of real wages and scarcities of food and goods. Agricultural and industrial production decreased, exports dropped, food supplies for the cities were inadequate, shortages of foreign exchange cut down on imports; city life became increasingly difficult and wages and salaries completely inadequate to live on. In 1955 the population of Kinshasa was 316,206, of which 111,724 were employed; in 1967 the population has risen to 901,251 but only 157,790 were employed. Thus in 1955 one worker was providing for 2.8 people, in 1967 for 5.7, while at the same time, between 1959 and 1967, there was a 50 percent drop in real salaries (Houyoux 1970, p. 99). This drastic situation made a woman's earning capacity even more essential for household support.

Women in rural areas had been the principal food suppliers and organizers of their households' subsistence. They had enjoyed relative economic independence in most societies, because once they had fed their household, they could sell the small surplus they produced and keep the proceeds to dispose of as they wished. The separate budgets maintained by men and women gave women a certain economic autonomy. As women moved into town, initially this advantage was taken away from them. But over time, economic exigency in the urban areas made women once again key food suppliers for their households as they acquired independent sources of income.

By 1963, LaFontaine could conclude that "for many Congolese women, independence of men is a new urban possibility" (LaFontaine 1974, p. 113). By 1965 the number of women in market trade had greatly increased, though only a few owned permanent boutiques. Bars, however, were the most profitable form of small scale enterprise for women. By this time also, a number of women were trading food between Kinshasa and Upper Zaire by boat, and to Lower Zaire by train or truck, and travelling by air to Kivu and Shaba (then Katanga) to buy more valuable goods such as bicycles, radios and cloth. These women operated outside the official economy. They did not use the banks or the savings banks, nor did they keep account books since most were semi-literate. Their starting capital was accumulated through the institution of *likelemba*, shortlived groups of three or four trusted friends with regular earnings, who agree to allocate a fixed sum every month to each member in turn. A few women with education and capital were even importers, a sector heavily dominated by men, chiefly foreigners (Comhaire-Sylvain 1968, pp. 182–88).

Joseph Houyoux, in both a preliminary study of sixty Kinshasa households in 1968 and a survey of 1,471 in 1969, discovered that expenses were greater than income for all socio-professional categories. Data from the preliminary study taken from twelve households of each of the five categories of unemployed, laborers, salaried workers, traders and managers provide a detailed picture of city life at the end of the sixties and show the crucial role that women played in household maintenance.

Houyoux's studies show that the primary source of income for the unemployed was the wife's petty trade. Women put what little money they could scrape together into buying sugar, manioc flour, fish, peanuts or flour wholesale and then sold them in small quantities, feeding the family on their meagre earnings. They also depended on gifts of money from kin or children established elsewhere, on earnings from occasional work by the husband and, in some cases, on prostitution by a wife or daughter. Some days they would go without any food at all (C. and J. Houyoux 1970, pp. 102–7).

The salaries of skilled and unskilled laborers in the twelve households under study averaged 1,474K[3] (ranging from 1,150K to 2,000K), including family allowances. On average they spent 877K more than their wages. The extra income came from the wife's commerce and from odd jobs undertaken by the husband. At the beginning of the month, eleven out of the twelve men gave their wives some money for commerce; wives then bought food for the family with their profits. These people too were often hungry. Salaried white collar workers from the lower levels of the administration and private enterprise averaged salaries of 2,638K (ranging from 1,670K to 3,200K), including family allowances. The husband never gave the wife enough money to feed the family; she always had to supplement her allowance. Ten out of the twelve wives bought a sack of rice, manioc or sugar with the money their husbands gave them and sold these foodstuffs in the market in small quantities. Two women had fields from which they fed their families; two were teachers and engaged in commerce during vacations. Household expenses surpassed salaries by 9.5 percent on average, and in general the difference was made up by the wife's trade. This gave a wife the "possibility of making her own living and [was] a symbol of emancipation" (C. and J. Houyoux 1970, pp. 107–115).

Traders' incomes were difficult to estimate because they did not keep accounts. With the exception of two cases their stock was not worth more than 2,000K. Six out of twelve of the traders' wives were in commerce independently of their husbands; three more worked in the fields, and three sold drinks from their houses. Managers consisted of school directors, officials, medical assistants and businessmen. Their salaries averaged 4,618K (ranging from 2,720K to 9,000K), but they also needed more to live; the 15 percent deficit was covered by their wives' activities. Two were in commerce, three worked in independent enterprise, and one cultivated a garden and sold the produce (C. and J. Houyoux 1970, pp. 115–19).

In the larger survey of 1969, 68 percent of household heads were salaried but their salaries represented only 55.3 percent of total household income (Houyoux 1973, p. 130). Sources of additional income for wage earners were various but 35 percent came from the commerce of wives. In general women bought food wholesale with money given them by their husbands and sold retail. There were four types of commerce: women purchased goods from the interior and sold them to market sellers, engaged in market trade in foodstuffs themselves, carried on commerce to the interior by boat, or had a little shop on their house lot to sell cigarettes, matches, jam and soap. Other sources of income were the husband's commerce or occasional jobs, the fraudulent abuse of his official position, or his transport or other enterprise. 4.1 percent of the wives of wage earners cultivated gardens in town and sold the produce; 7 percent of the households in the survey received income from rents. For the unemployed the most important source of income was wives' commerce; the retired, the elderly and students supplemented pensions or scholarships with commerce, rents and gifts from kin. In all categories some women earned income from prostitution (C. and J. Houyoux 1970, pp. 242–247).

By the end of the 1960s, therefore, women had begun to achieve more economic independence and autonomy than they had had during the colonial period. Thereafter, some of them greatly expanded their activities in the proliferating activities of the second economy, in some cases even investing accumulated capital in substantial businesses in the official economy.

The Second Economy in the 1970s and 1980s

The enormous expansion of the second, unmeasured or unrecorded, economy in the 1970s and 1980s is attributable to Zaire's deepening economic crisis, which has resulted in acute shortages of food, fuel and manufactured goods and to the steady decline of the administrative capacity of the state. Unlike the colonial state, the post-independence state has been generally incapable of enforcing price controls, regulations on obligatory cultivation and licensing requirements, and also of preventing smuggling of major export commodities. Such weakening of the state and expansion of the second economy involves a decline in the effectiveness of the mechanisms of male domination and control (as Jane Parpart (1987) also points out).

The second economy consists of a highly organized system of income-generating activities that deprive the state of taxation and foreign exchange; they are unrecorded in official figures and left out of official reports. Some of these activities are illegal, others are legitimate in themselves but carried out in a manner that avoids taxation. This sector of the economy consists primarily of clandestine gold and diamond mining and ivory poaching; smuggling and theft; barter and other forms of unlicensed trade; speculation, hoarding and middleman activity; and bribery, corruption and embezzlement (for details see J. MacGaffey 1987, ch. 5).

As mentioned above, by the mid-sixties some Zairian women were becoming successful and wealthy in large scale trading operations, primarily in the second economy. Since trade in this economy is carried on without licenses, does not involve use of the banks, nor require account books or much if any use of French, a woman does not need either her husband's permission or much education to practise it. She operates outside the control of the state by evading income taxes and licensing requirements. Women with large scale business operations have commercial licenses, but do not necessarily keep accounts or declare all their activities and income. By 1979 many women had been able to expand their trading activities, and some had accumulated considerable wealth, invested in business enterprise in the official economy, and improved their class position and lifestyle.

Some women are active in the second economy in the lucrative ivory smuggling trade;[4] others engage in speculation, shipping goods from a part of the country where they are scarce to another where they are in short supply, and charging a high markup. Others use personal connections to acquire the scarce foreign exchange needed to run their own profitable importing business. It is impossible to quantify the exact scale of individual operations, just how many women carry out these activities or the amount of their incomes. As one man put it, "they will never reveal their earnings and details of their trade, they *do not even want their husbands or families to find out what they do.*" This is exactly the point. They do not want to be open about their dealings and thus be part of the official economy; they have escaped male constraints by moving into the unrecorded activities of the second economy. Many such women are unmarried, divorced or widowed; their freedom from wifely obligations, as well as the exclusion from professional employment that all women suffer, simply makes it easier for them to operate in the second economy.

One lucrative form of trade in which women specialize is in wax print cloth. It is sold in lengths for the standard women's costume of wraparound skirt and blouse, and ranges from a cheap, locally produced print to the highest quality imported Dutch Java, which cost as much as Z1,200 in 1980.[5] In one instance a woman seller of these prints reportedly grossed Z45,000 in a single morning (Vwakyanakazi 1982, p. 226). Another profitable commerce is in foodstuffs from the interior. Fish, rice and beans are shipped from Kisangani down the Zaire river to Kinshasa. Some businesswomen in Kisangani ship as many as 100 to 200 sacks of beans a month; other women come up from Kinshasa by plane and buy whole truckloads at a time to ship downriver. This trade is an important item in the food supply of the capital. The major source of beans, a staple food, is in Kivu, the easternmost region of Zaire; Nande traders import them by road from Kivu to Kisangani where they are transhipped to river boats. Fish is another lucrative commodity. A Greek wholesaler in Kisangani said woman traders regularly bought Z30,000–50,000 ($6–8,000) of smoked fish from him at a time to ship down to sell in Kinshasa. Some women have thus moved into commerce on a very considerable scale.

In the seventies and eighties, therefore, women's opportunities expanded further, permitting them to do far more than merely help their families to survive. Some of them managed to greatly increase the scale of their activities. They moved from petty trade into extensive and lucrative enterprise, some of it invested in the official economy. A number of them became renowned for their success and wealth; in Kinshasa by 1979 some women were millionaires, others had bank accounts of Z100,000–200,000. The possibility for independence and autonomy for urban women observed by LaFontaine had become reality on a scale far beyond anything that existed in the sixties.

In 1980, nine women owned substantial businesses in Kisangani, Upper Zaire, one of Zaire's three principal cities.[6] They were engaged in retail and semi-wholesale trade, in travelling commerce, and in exporting and importing goods to and from Kinshasa or to the interior, and were investing in expanding their businesses, in plantations and in real estate. Recognition of the success of women in business was reflected in the election of a Kisangani businesswoman to the regional committee of the Chamber of Commerce in 1980.

I have argued elsewhere (J. MacGaffey 1987) that these Kisangani women, unlike the politicians' wives who own businesses as "fronts" for their husbands, are part of an emerging indigenous capitalist class, small as yet, but distinct from the "political aristocracy" and its parasitic form of capitalism. This new commercial middle class is composed of substantial business owners who do not hold political position and who are relatively independent of politics. They manage their enterprises in rational capitalist fashion and reinvest profits to expand their businesses.

Women make up 28 percent of the independent business owners in Kisangani. Women also make up nearly one quarter of the retailers with fixed stores in the central commercial and administrative zone of the city, and they dominate market trade. The nine women who were substantial business owners in 1980 had each started out on a small scale in the way described for women in Kinshasa in the 1960s. Four had begun in petty trade, one by owning a bar, one by working as a teacher and trading on the side, and one by working as a prostitute; only two had been given initial capital by their husbands. Particular events and circumstances had provided the opportunities which led to their later success.

I have described in more detail elsewhere the methods through which these women built up large scale enterprises (J. MacGaffey 1987, ch. 7). Their individual case histories reveal the impact of the wider political and economic context: the political troubles following independence, the Zairianization and retrocession decrees indigenizing foreign capital, and the expansion of the second economy. The scale of women's business activities is such that "Mamabenzi" is now a popular category in Kinshasa, the counterpart for women of the East African term "Wabenzi," referring to affluent men owning Mercedes Benz cars.

Men as an Economic Resource

One way women get access to resources needed for participation in the second economy or other arenas is to manipulate sexual relations. LaFontaine writes that in Kinshasa during 1962–63, *femmes libres* (a term generally translated as prostitute but used to refer to prostitutes, mistresses, concubines, temporary companions and single women indiscriminately) were

> . . . not bound by the rules of wifely modesty that restrain the behaviour of married women. . . . Since they have more contact with the masculine world they often have a much greater sophistication than 'respectable' women. Many of them learn good French. . . . While most are uneducated and few are the chic cosmopolitaines of the ideal, . . . in enacting the role of companions, prostitutes are also 'free' of traditional stereotypes of feminine subordination to a man (LaFontaine 1974, p. 96).

Men are ambivalent about *femmes libres* and their independent behavior: they admire them but forbid such actions in their wives. For women there is thus a choice between two contrasting roles. An essential feature of love affairs is the gifts given to a woman by her lover, usually clothes and personal finery, but also cash or expensive household items or even a car. Successful courtesans are known as *vedettes* ("stars"); to become rich and famous through having many important lovers is one kind of feminine ambition, one ideal type of modern success (LaFontaine 1974, pp. 97–98). "*Femmes libres* dispose of themselves and their resources in their own interests, create and manage their own relationships and do not submit to control of these by men, either as kinsmen or husbands" (LaFontaine 1974, p. 111). Some women turn such relationships to their advantage in getting goods or foreign exchange for their trading activities.

Carmel Dinan has emphasized that in Ghana the urban environment offers a greater variety of economic roles and opportunities for women than the traditional order and that sexuality, since it is not bound up with sin in religious belief systems, as in Europe, nor with the refinements of romanticism, can legitimately be viewed more objectively and instrumentally. In Accra, *machismo* was expressed in men's ability to attract and maintain attractive "girlfriends."

> Such a situation provided a very fertile area for female manipulation, especially because women have managed to preserve continuity with the traditional pattern: in return for their sexual services, the men had to assume responsibility for their main financial outlays—rent, food and clothing. Their sexual roles were, consequently, of considerable economic potential (Dinan 1983, pp. 353–54).

These "sugar daddy/gold-digger" relationships were exploitative on both sides. Women offered straight-forward economic reasons to account for them and perceived them as a direct exchange of services. Women in such relationships, as well as those in white-collar employment, were accumulating capital and investing in trading items for the time when they planned to become full-time businesswomen; salaried employment was not their final goal (ibid., pp. 356, 358, 360). In Zaire too women have made use of the economic potential of their sexual roles, for survival, for success and autonomy, and for accumulating wealth to invest in business enterprise. Wilson asserts that an increasingly popular strategy is to refuse to marry; ambitious women prefer to live with a man without marrying him, or as the girl friends of more than one wealthy man, in this way hoping, though not without risk, to achieve economic independence (Wilson 1982, pp. 166–67).

Women have thus found opportunities and strategies for becoming autonomous and, in some cases, extremely wealthy, sometimes independently of men by operating in arenas beyond their control and sometimes by using them to gain access to resources.

Changes in Family and Kinship Structure

The greater autonomy and independence of women in the towns than in rural society, and the wealth achieved by some of them, is affecting the organization of family and kinship. Jane Guyer has emphasized the importance in modern Africa of the intervention of the state for changes in marriage, and in kinship and property relations (Guyer 1984, p. 80). In Zaire, as we have shown, the colonial and post-colonial state has intervened in all these areas and has redefined the position of women and structured men's control over them in the process of urbanization. However, women's extensive participation and accumulation of wealth in the second economy has provided new directions for recent changes in kinship and marriage, this time brought about by activities outside state control and evading male domination. In this way some women have managed to escape from the unfavorable position to which they were relegated in urban society.

In 1962–63 in Kinshasa, LaFontaine notes that "the economic independence of women has meant in some cases that kinswomen are becoming as important as kinsmen." She gives an example of a woman owner of a parcelle who was temporarily supporting four men from her village in search of work. She also found that other women supported dependents other than their children and themselves and that a larger proportion of single women supported dependents than did single men. She concluded that the single woman was often better off than the lower-paid single man and more conscientious in fulfilling kinship obligations (LaFontaine 1970, pp. 139–40). Kisangani in 1980 furnished an example of a prominent businesswoman assuming the role generally undertaken by a senior kinsman. This woman supported her deceased sister's five children and paid for them all to go to a mission boarding school. She regularly received kinsfolk in trouble, spending afternoons in an annex to her house resolving problems and giving out money for sudden large expenses, such as medical care. When an uncle died in a town in the interior, she organized and presided over the wake for all kin resident in Kisangani.

LaFontaine also found that some independent women appeared to cut themselves off from contact with rural kin when they were established in town, more so than did men. This apparently reflected the advantages that town life could offer women as compared to village life, where they had no claims to leadership and could not attain the prestige they could in towns, so that they had less incentive to maintain rural kin ties (LaFontaine 1970, p. 151). Kin in town were more important to them, however. In 1965, Comhaire-Sylvain found that many of the women in her survey felt more obliged to give presents to their own kin than to share their salary with their husbands (1968, p. 177), thus reflecting the greater security offered by the support of kinsfolk than by an often unstable marriage.

The trend in the city, however, is towards the restricted family and the declining importance of the descent groups that structured traditional rural society. As individuals acquire businesses and property in the city, they want to pass them on to their children, and are reluctant to disperse holdings among a wide range of kin, as is customary in traditional society. The right to a parcelle is a valuable inheritance and has given rise to inheritance disputes; decisions in cases settling these disputes have resulted in a new body of common law for the city, a direct outcome of the changes taking place in family structure in the urban environment (LaFontaine 1970, pp. 60, 80).

As Houyoux points out, the increasing economic role of women in urban areas has various consequences for the nuclear family. Commerce gives independence to a woman and makes her freer in her relationships but is a source of new conflicts within the household. A woman's absences to pursue her trade are problematic for child raising in the new restricted family, which lacks the wide range of kin who would have formerly participated in child care (Houyoux 1973, p. 245). One area of conflict is the restructuring of the control and distribution of household income as the economic independence of women increases. The proposed new family code attempts to reassert male control over a woman's potential independence: "Commentaries attached to the Articles make clear that the goal of 'consolidation of the nuclear family' means mostly the accumulation of capital under the control of the husband" (W. MacGaffey 1982, p. 94). So the struggle continues.

As Lourdes Bénería asserts, the subordination of women is related to
the basic economic and political structures of society, and "one of the
dimensions of women's oppression is the existence of mechanisms of
exploitation that feed on and accentuate inequalities related to class and
gender" (Bénería 1982, p. xvi). In the Belgian Congo the colonial state
organized the penetration of capitalism and a class structure which supported
the interests of metropolitan capital. In this process women's position in
society worsened. After independence, upward mobility into the new
dominant class depended on education and position in the state, from which
women, with few exceptions, were excluded. But changes in the basic
political and economic structures, in particular the weakening of the
administrative capacity of the state, the indigenization of foreign capital,
the deterioration of the formal economy, and shifts in the class structure
have recently brought about considerable improvement for at least some
urban women and a change in the expected role of women in general in
urban society.

Notes

1. Keith Hart (1973) first distinguished the "informal sector" of the economy in
Ghana, and introduced the distinction betwen formal and informal income opportunities. The literature on this subject since Hart, however, has dealt chiefly with
the small-scale activities of the urban poor (for a summary see Moser 1978), omitting
activities such as smuggling, speculation and corruption, included by Hart. These
are often very large scale and carried out by those at the top level of society. Other
terms for this sector of the economy are the second, parallel, underground or black
economy. When it is extensive relative to the formal economy, and wealth is
accumulated in it on a very large scale, the connection between the state and class
formation becomes questionable (Kasfir 1983; J. MacGaffey 1987).

2. Kenneth Little defines prostitute as a woman "whose livelihood over a period
of time depends wholly on the sale of sexual services and whose relations with
customers does not extend beyond the sexual act." He points out that in West
Africa prostitution overlaps with other kinds of relationships which may be quasi-
uxorial, also that women may earn money by this means only periodically (Little
1973, pp. 84–87). Likewise, Luise White describes various kinds of domestic services
offered by prostitutes in Nairobi (White 1983; this volume).

3. In 1967 the Congololese franc was replaced by the zaire (Z) divided into 100
makuta (K) . Z1 = US $2.

4. In March 1980, officials in Athens intercepted four cases containing 1,200
kilos of ivory exported undeclared from Zaire; a further three and a half tons were
seized from a large truck in Zaire. Both belonged to a woman apparently acting as
agent for an influential politician in Kinshasa (*Boyoma*, Kisangani's daily paper, 27
March 1980). Sale of ivory was prohibited in April 1979 but it sold clandestinely
in Kisangani 1979–80 for Z200 a kilo, so the total value of the confiscated ivory
was around Z940,000 (approximately US $156,600).

5. Z1 = US $0.34 Feb. 1980. The black market and realistic rate was Z1 = US $0.20.

6. I did 10 months anthropological field research in Kisangani, from Sept. 1979 to June 1980, as a research associate of the Centre de Recherches Interdisciplinaires pour le Developpement de l'Education (CRIDE). My research was aided by a Grant-in-Aid of Research from Sigma Xi, the Scientific Research Society. I used the usual anthropological methods of participant observation and informal interviewing.

References

Beneria Lourdes. (1979) "Reproduction, Production and the Sexual Division of Labor," *Cambridge Journal of Economics* 3, 3 (Sept.): 203–25.

———. ed. (1982) *Women and Development: the Sexual Division of Labor in Rural Societies.* New York: Praeger.

Boserup, Ester. (1970) *Woman's Role in Economic Development.* New York: St. Martin's Press.

Collier, Jane Fishburn. (1974) "Women in Politics," pp. 89–97 in M.Z. Rosaldo and L. Lamphere, eds., *Women, Culture and Society.* Stanford: Standford University Press.

Comhaire-Sylvain, Suzanne. (1968) *Femmes de Kinshasa: Hier et Aujourd'hui.* Paris: Mouton.

Dinan, Carmel. (1983) "Sugar Daddies and Gold-Diggers: the White Collar Single Women in Accra," in Christine Oppong, ed., *Female and Male in West Africa.* London: George Allen and Unwin.

Gould, Terri F. (1978) "Value Conflict and Development: the Struggle of the Professional Zairian Woman," *Journal of Modern African Studies* 16, 1: 133–139.

Guyer, Jane. (1984) *Family and Farm in Southern Cameroon.* African Research Studies, 15. Boston: Boston University African Studies Center.

Hart, Keith. (1973) "Informal Income Opportunities and Urban Employment in Ghana," *Journal of Modern African Studies* 11, 1: 61–89.

Houyoux, C. and J. (1970) "Les Conditions de Vie dans Soixante Familles a Kinshasa," *Cahiers Economiques et Sociaux* 1: 99–132.

Houyoux, Joseph. (1973) *Budgets Menagers, Nutrition et Mode de Vie a Kinshasa.* Kinshasa: Presse Universitaire de Zaire.

Kasfir, Nelson. (1983) "State, *Magendo* and Class Formation in Uganda," *Journal of Commonwealth and Comparative Studies* 21, 3: 84–103.

LaFontaine, Jean S. (1970) *City Politics: a Study of Leopoldville 1962–63.* Cambridge: Cambridge University Press.

———. (1974) "The Free Women of Kinshasa," in J. Davis, ed., *Choice and Change: Essays in Honor of Lucy Mair.* New York: Humanities Press.

Little, Kenneth. (1973) *African Women in Towns.* London: Cambridge University Press.

MacGaffey, Janet. (1987) *Entrepreneurs and Parasites: The Struggle for Indigenous Capitalism in Zaire.* Cambridge: Cambridge University Press.

MacGaffey, Wyatt. (1982) "The Policy of National Integration in Zaire," *Journal of Modern African Studies* 20, 1: 87–105.

176   *Evading Male Control*

Manwana Mungongo. (1982) "Les Droits de la Femme Travailleuse au Zaire," *Zaire Afrique* 163: 73–82.

Mbilinyi, Marjorie. (1985) "The Changing Position of women in the African Labor Force," pp. 170–186 in Timothy M. Shaw and Olajide Aluko, eds., *Africa Projected*. New York: St. Martin's Press.

Moser, Caroline. (1978) "Informal Sector or petty Commodity Production: Dualism or Dependence in Urban Development?" *World Development* 9/10: 1041–1064.

Parpart, Jane. (1987) "Women and the State in Africa," in Naomi Chazan and Donald Rothchild, eds. *The Precarious Balance: State and Society in Africa*. Boulder, Colorado: Westview Press.

Pons, Valdo G. (1969) *Stanleyville: an African Community under Belgian Administration*. London: Oxford University Press:

Rosaldo, Michelle Zimbalist and Louise Lamphere, eds. (1974) *Woman, Culture and Society*. Stanford: Stanford University Press.

Schwartz, Alf. (1972) "Illusion d'une Emancipation et Alienation Reelle de l'ouvriere Zairoise," *Canadian Journal of African Studies* 6, 2: 183–212.

Vwakyanakazi, Mukohya. (1982) *African Traders in Butembo, Eastern Zaire (1960–1980): a Case Study of Informal Entrepreneurship in a Cultural Context of Central Africa*. Ph.D. Dissertation, University of Wisconsin, Madison.

White, Luise. (1983) "A Colonial State and an African Petty Bourgeoisie: Prostitution, Property and Class Struggle in Nairobi 1936–1940," pp. 167–194 in Frederick Cooper, ed., *The Struggle for the City*. Beverly Hills: Sage.

Wilson, Francille Rusan. (1982) "Reinventing the Past and Circumscribing the Future: Authenticité and the Negative Image of Women's Work in Zaire," pp. 153–170 in Edna G. Bay, ed., *Women and Work in Africa*. Boulder, Colorado: Westview Press.

Yates, Barbara. (1982) "Colonialism, Education and Work: Sex Differentiation in Colonial Zaire," in Edna G. Bay, ed., *Women and Work in Africa*. Boulder, Colorado: Westview Press.

9

The Middle-Class Family in Kenya: Changes in Gender Relations

Sharon B. Stichter

In the introduction to a recent special issue of the *Journal of African History*, the editors report their concern that the history of the family, which has become so lively and important an area of study in Europe and America since the 1960's, was being almost totally neglected in Africa (Marks and Rathbone 1983, p. 145). It is indeed surprising that contemporary Africanists scholars have paid so little attention to the transformation of the family over time. One reason may be precisely the rich tradition of anthropological kinship studies which has for so long directed our attention to the static classificatory aspects of kinship rather than to the politics of kinship or its processes of change. Now, however, a newer more promising approach sees kinship as cultural ideology conditioned by social structural factors such as the emotions and material interests of members of domestic units (Medick and Sabean, 1984; Poster, 1978).

The typical pitfalls of family history obtain in Africa as much as anywhere else. There is for example the temptation to romanticize the past, which is all the more appealing to those who wish to support Africa's struggle to assert and preserve its identity in the face of western economic and cultural onslaughts. Nostalgia for the true "African family" threatens to prevent scholars from grasping the amount of change that has taken place. But nostalgia is precisely a reaction to change. One study of Malawi observes that paradoxically "the emphasis placed on the clan unit can be explained partly by its present-day decline . . . " (Vaughan, 1983, p. 279).

Another pervasive problem, of central concern in this paper, is the tendency to see household, kin and family from a male viewpoint. To focus

The author wishes to thank Jane Leserman for statistical consultation and Leon Margelis for programming assistance.

177

on the "politics of the family" means to take seriously the two chief axes of conflict: gender and generation. Most of the work of reproduction, and much of that of production, is done by women in the family: marriage and kinship rules establish, by and large, male rights to the products of women's labor. It follows that to comprehend the functioning of the domestic unit an essential step is to disaggregate it, to analyse the interests, emotions and strategies of its individual members. Since nearly all known family systems are to some extent patriarchal, analysing the strategies and consciousness of women and juniors will give us what has been missing from family history: the "view from below," the "categories/non-categories of the dominated," the sources of resistance. The deconstruction of the domestic unit may actually be easier to accomplish in sub-Saharan Africa than in some other parts of the world, given this area's historic emphasis on the division by sex of labor, property rights, and spheres of influence.

The avoidance of male and elder bias obviously requires an at least temporary shift in the level of analysis from the household or kin unit level to that of the individual. There is always a tension between a focus on the individual and a focus on social structures. There is a well-established view that the social whole is inevitably more than the sum of its parts, or that some individual interests must always be subordinated to the social good. More specifically, some argue that the western feminist emphasis on the individual does not recognize the central importance of household and marriage relations for the economic survival of most women in the Third World (e.g. Sharma, 1986). Yet one could not even arrive at such conclusions without asking the prior question of what the individual person's or women's interests are in a given situation. Analytically, we cannot continue to assume that households, conjugal units, or lineages have strategies, interests or "utility functions" into which all members have had equal input or from which they derive equal benefit.

This paper addresses the changes in women's position in the family that have come about as a result of the creation of a middle class in Kenya. It focuses on the changes in male-female relations in (1) material appropriation and exchange within the conjugal unit; (2) conjugal power and decision-making; and (3) domestic labor. The underlying comparative question is whether more egalitarian and more "joint" relations are coming into being in the domestic domain, such as are said to exist in contemporary European and American middle-class families. If they are, this would constitute a change from the gender-divided yet patriarchal African family of the past. Changes in gender relations in the family can be seen as part of the broader question of whether a transition to the western "bourgeois" family is taking place in urban Africa.

Middle-Class Families in Nairobi

This paper draws on the results of a survey of 317 families living in two fairly new housing estates on the edge of Nairobi, Buru Buru Phases 1 and 2, and Umoja.[1] Ten *percent* and 5 *percent* systematic random samples of households were chosen in Buru Buru and Umoja respectively, and both the wife (or wives) and the husband were interviewed. The survey was carried out in July and August, 1979. Both these estates are very large; the two phases of Buru Buru together have about 1,897 units; Umoja at the time of interviewing had 2,903 units. Both estates were built with external development assistance, then sold to Africans who met the minimum income levels needed to qualify for the mortgage. The style of housing varies: Buru Buru is two-story wood frame construction with three to five rooms and a private parking space, while Umoja is one-story concrete block construction with two-room units. The cost of the housing varies accordingly, Umoja being much lower cost.

Our sample is representative of Nairobi's small but growing African middle-class, employed in commerce, services, light industry and government. White-collar workers predominate in the sample (Table 1). Among the men, the largest category is white-collar workers, with a good percentage of technical, professional and managerial employees as well. The only estate with a significant proportion of male manual workers is Umoja.

A great many middle-class conjugal units in urban Kenya today are two-earner units. Wives usually contribute an important share of the household income, either through wage employment or self-employment. Nearly 70 percent of the women interviewed were active in the labor force (Table 2). In this sample the majority were in white-collar jobs, most frequently typists, secretaries or clerks. A third of them were in professional-level jobs, almost all primary school teachers or nurses. Hardly any women held managerial positions, in contrast to the men. A small number were self-employed, most of these being in Umoja.

The median household income in the sample was L1740. The median income in Nairobi for individuals in 1979 was estimated at L1200 to L1300 per annum; our sample thus approximated the median. But the income range was substantial, and the model income varied by estate, from a high of L3000–L3599 per annum in Buru Buru 2 to L1200–L1799 in Buru Buru 1, to L600–L1199 in Umoja (Table 3). Because of a shortage of good housing in the city, there are a fair number of high-income households living in Buru Buru who may be hoping to move to more elite sections. In Umoja, many of the inhabitants are young, and hoping to move to better quarters.

Education is one of the defining attributes of the African middle class. In the sample, over half of both men and women had completed Form 4

TABLE 1   OCCUPATIONS (percentages)

	BB2	BB1	Umoja	Total
WOMEN				
Manual	8.1	1.7	3.8	4.7
White Collar	51.6	63.2	75.0	62.5
Professional	40.3	35.1	21.2	32.7
	----	----	----	----
	100.0	100.0	100.0	99.9
MEN				
Manual	2.9	2.9	14.2	7.9
Technical	19.1	22.1	18.9	19.8
White Collar	35.3	35.3	49.0	41.3
Professional	17.6	25.0	13.2	17.8
Managerial, Administrative	25.0	14.7	4.7	13.2
	----	----	----	----
	99.9	100.0	100.0	100.0

Source: Author's survey
Deviations from 100 due to rounding

TABLE 2   EMPLOYMENT STATUS - WOMEN   (percentages)

	BB2	BB1	Umoja	Total
Employed Full Time	83.1	75.3	46.3	64.7
In School or Training	1.3	1.3	3.3	2.2
Unemployed, Looking for Work	5.2	6.5	18.2	11.3
Unemployed, Not Looking or No Information	9.1	15.6	28.1	19.3
(Total Unemployed)	(14.3)	(22.1)	(46.3)	(30.6)
Self-Employed	1.3	1.3	4.1	2.5
	-----	-----	-----	-----
	100.0	100.0	100.0	100.0

Source: Author's survey

TABLE 3   TOTAL FAMILY INCOME   (percentages)

Kenya L per annum/ Kenya Shs. per month	BB2	BB1	Umoja	Total
0-119/0-199	--	--	--	--
120-359/200-599	--	--	--	--
360-599/600-999	0	2.9	3.8	2.5
600-1199/1000-1999	3.0	7.4	34.9	18.3
1200-1799/2000-2999	7.5	23.5	22.6	18.7
1800-2399/3000-3999	14.9	16.2	20.8	17.8
2400-2999/4000-4999	19.4	17.6	7.5	13.7
3000-3599/5000-5999	20.9	17.6	5.7	13.3
3600-4799/6000-7999	16.4	11.8	3.8	9.5
4800-5999/8000-9999	11.9	0	.9	3.7
6000+ / 10,000	6.0	2.9	0	2.5
	100.0	99.9	100.0	100.0

Source: Author's survey
Deviations from 100 due to rounding

secondary education. However, only 60 *percent* of the women had completed secondary school, as contrasted to 85 *percent* of the men (Table 4). Differences among the estates in education parallel those in income, although Umoja does not differ so markedly from Buru Buru in this respect, partly because Umoja residents are on the average ten years younger and either they have not yet reached their maximum earning capacity or they are finding that education does not so easily translate into income as it did a decade earlier. Still, among the women, differences are evident: a greater percentage in Buru Buru 2 than in Buru Buru 1, and in Buru Buru 1 than in Umoja, had completed some post-secondary training, and a smaller proportion in Buru Buru 2 than in the other two areas had completed only primary school.

The unemployment rate among women was quite high—30.6%. It was highest, 46.3% in the lower-income estate, Umoja (Table 3). Education, as noted, followed a different pattern as between estates, so that education alone probably does not account for whether or not a woman is employed. However, the zero-order correlation between women's employment and their education was $r = .44$, significant at the .01 level.

Household structure affected women's employment in that virtually all of the unmarried, divorced and widowed women interviewed were employed. Evidently they had to be in order to afford to live in the estates, though some may have been receiving some support from boyfriends or absent

TABLE 4   EDUCATION   (percentages)

WOMEN

	BB2	BB1	Umoja	Total
None	1.2	1.3	3.3	2.2
Standard 1-4	1.3	1.3	3.3	2.2
Standard 5-8	17.7	28.2	23.1	23.0
Form 1-2	12.6	11.5	12.4	12.2
Form 3-4	50.6	47.4	49.6	49.3
Over Form 4, Technical	16.5	10.3	7.4	10.8
University	0	0	.8	.4
	100.0	100.0	99.9	100.1

MEN

	BB2	BB1	Umoja	Total
None	0	0	0	0
Standard 1-4	0	1.4	.9	.8
Standard 5-8	5.9	5.8	9.6	7.6
Form 1-2	5.9	7.2	3.5	5.2
Form 3-4	48.5	43.5	53.5	49.4
Over Form 4, Technical	38.2	42.0	30.7	35.9
University	1.5	0	1.8	1.2
	100.0	99.9	100.0	100.1

Source: Author's survey
Deviations from 100 due to rounding

husbands. The percentage of female-headed households (categories 1 and 5 in Table 5) was fairly high in the estates: 10.3% overall, 8.6% in Buru Buru 2, 11% in Buru Buru 1, and 11% in Umoja. In Buru Buru 2, the female-headed households were nearly all married women whose husbands were temporarily away, whereas in Umoja they tended to be single, separated or widowed women and as such, much more at risk financially. Since the questions in this study focused on the position of women within formal marital relationships, subsequent analysis is based on the sub-sample of married households only.

Interestingly, the development cycle of the family, reflected in the ages of the children, had virtually no effect on whether the woman was employed. Of all those employed, 77.7% had one or more children under six years of age, which exceeded the sample percentage of women with children in that age bracket. This pattern of uninterrupted labor force participation

TABLE 5   HOUSEHOLDS: MARITAL AND RESIDENTIAL CHARACTERISTICS
          (percentages)

	BB2	BB1	Umoja	Total
Single, separated, divorced, widowed, Female	1.2	7.3	8.0	6.0
Single, separated, divorced, widowed, Male	1.2	0	.7	.7
Married, One Wife, Both live there	84.0	78.0	69.3	75.7
Married, One Wife, Wife lives elsewhere	0	6.1	14.6	8.3
Married, One Wife, Husband lives elsewhere	7.4	3.7	3.0	4.3
Married, Two or more wives	6.2	2.4	3.6	4.0
Unmarried couple	0	2.4	.7	1.0
	100.0	99.9	99.9	100.0

Source: Author's survey
Deviations from 100 due to rounding

among women is similar to that emerging in highly industrialised nations today, and contrasts with Asia and Latin America today (as well as with the United States in the 1940's and 1950's) where women often drop out, or are pressured out, of wage employment when their children are young.

Ethnicity was also significantly related to women's employment. The female sample as a whole was 46.8% Kikuyu, Embu and Meru; 17.1% Luyia; 15.7% Luo, 10.2% Kamba, and 10.2% other. The ethnic distribution for men was similar. Kikuyu, Embu and Meru women had the highest rate of employment—70.8%—followed by Kamba, 63.3%, Luyia, 56%, and Luo 41.3%. Given the average rate of employment among women of 64%, Luo women were severely underemployed. Fifty-two *percent* of Luo women were unemployed, as contrasted to 24.8% of Kikuyu, Embu and Meru. This fact could result from discrimination against Luo in hiring markets. However, it is even more likely to be related to the greater persistence of patriarchal controls over women among the Luo. Differences in educational qualifications between Luo and Kikuyu women were also evident. Clearly, a combination of labor market supply and demand factors is operating.

Aspects Of Household Structure

Presentation of survey results in terms of *household* attributes can be misleading in Africa. As Jane Guyer has recently argued, moving to a household kind of analysis implicitly assumes that a transition to the "conjugal state" has already taken place, when in fact it is that very transition which is in question (Guyer, 1981, p. 97). Jack Goody has convincingly set forth a broad contrast between the traditional systems of inheritance and other aspects of kin and domestic relations in subSaharan African societies, as compared to those in Eurasia. Africa is said to be characterized by homogeneous or sex-linked inheritance, bridewealth and polygyny, these being compatible with low-productivity hoe agriculture and a relative lack of economic differentiation. Eurasia, on the other hand, exhibits "diverging devolution" through the joint or conjugal estate, dowry as the principal form of female inheritance, and monogamy, these features being linked to plough agriculture, the wheel, and class differentiation (Goody, 1976). Reviewing recent studies of the African household and lineage, Guyer affirms the continuing relevance in rural areas of Goody's observations: "African kinship is not necessarily developing in the direction of the corporate household more characteristic of classic peasantries" (Guyer, 1981, p. 104). But the chief inadequacy of Goody's approach, as Ann Whitehead has pointed out, is that it does not include reproductive rights in its account of the transmission of property through inheritance. Thus it conveys a false sense of symmetry in men's and women's property inheritance, when in fact they are completely different. The more production requires people, as it does in the labor-intensive African farming systems, the more will a major aspect of property relations consist of rights in people (Whitehead, 1977; 1984).

All this still leaves open the question of whether among urban wage-earning families, whose major form of material property is the wage or salary income, a transition toward conjugality might be taking place. More relevant to these cases is William Goode's classic hypothesis that "modernization" inevitably leads to the emergence of the nuclear family (Goode, 1964; applied to Africa, see Lloyd, 1967). The nuclear family as a theoretical construct combines the notions of small household size, emotional closeness and privacy, separation from extended kin, young people's freedom from control by their elders, and increased egalitarianism between husband and wife. As Scott and Tilly point out, this family form emerged among the European middle classes in the 19th century and did not diffuse to the working class until well into the 20th century (Scott and Tilly, 1974). In their terms, there was an uneven movement from the "family economy" of the countryside, to the "family wage economy" of the urban artisans and workers, among whom women and wives continued to perform much wage work and home production, to the early 20th century middle and working-class "family consumer economy" in which the wife participates less in the labor force, relying on the husband's wage and managing the increasing amounts of household consumption (Tilly and Scott, 1978).

The difficulty in applying Tilly and Scott's conception to Africa is that the notion of "family" employed is that of the European conjugal household, with its assumption of pooled, shared resources. The African middle class unit is not a "family consumer economy" even though many commodities are purchased and consumed. First, the wives have not withdrawn from the labor force to live from their husband's wage. Second, the reason they have not has probably less to do with "family survival strategies" than with the weak sense of conjugality in Africa. If, as Goody reminds us, the woman cannot rely on her husband to pass on all or most of his income and property to her and her children, she is faced with the necessity, long established in African customary norms, of providing for herself. The survival and prosperity of the mother-child sub-unit, the "hearth-hold" in Felicia Ekejiuba's term (1984), is more likely to be the focus of women's efforts than is the conjugal unit.

In our sample the modal household size was 7 in Buru Buru 2, 8 in Buru Buru 1, and 5 in Umoja. This is smaller than in African rural areas, but larger than among the European middle class. Usually, there were at least two other persons resident in addition to the husband, wife and children: one usually the maid or "housegirl," who might be a distant relative, the other often another relative who was a temporary guest. This distinctly African household composition was not well provided for in the architectural design of the housing estates. Buru Buru in particular appeared to be designed for the idealized western nuclear family of two children. The units had small rooms, divided into living room, kitchen, and one or two bedrooms. High fences around the yards further emphasized small-family privacy. In spite of the structure, guests, housegirls and children slept anywhere possible—in halls, on living room sofas, even in the kitchen.

The average number of live births per woman in Buru Buru 2 was 3.18, whereas in Buru Buru 1 it was 3.43. The age distribution of women in these two estates was similar. The difference between the two figures, and the low level generally, illustrate the impact of female education and employment on fertility. The mean number of live births per woman in Umoja was 2.42, reflecting the younger age of the women there. The percentage of children living at home with their mother was about 90%. Children were not being frequently transferred around or fostered; however, older children might be sent to secondary school in the rural areas, sometimes staying with relatives. A small but noticeable proportion (8%) of the wives had their primary residence elsewhere, mainly in rural areas (Table 5). This arrangement was more common in the lower-income estate.

The legal and actual persistence of polygyny in Africa undercuts the formation of a "conjugal estate" or "nuclear family." The fact that any African marriage is at least potentially polygynous dilutes the wife's claims on her husband's income in comparison to those she might make under monogamy. The rate of reported polygyny in this sample was only 47% (Table 5), much below, for example, the levels found by Parkin in 1968 in the lower income estate of Kaloleni, where the rates varied from 7% for the Kikuyu to 33% for the Luo (Parkin, 1978, p. 45). In our sample, being the only wife was significantly positively correlated with the education level of the wife (but not that of the husband), and with being Kikuyu. It was negatively correlated with large age and education differences between husband and wife, with the age of the husband, and with being a Luo wife or husband. Having only one wife was noticeably *not* correlated with either the education of the husband or with household income; in face, the polygyny rate was highest in the higher-income estate (Table 5). This suggests that increasing male incomes and male education levels may not lead to a decline in polygyny, but that increasing female education levels may.

The rate of formalised polygyny only scratches the surface, however, it is impossible to assess the rate of "disguised polygyny," the practice whereby a man keeps a mistress or "outside wife" in a separate apartment, often without the knowledge of his first wife. Rumor had it that some of the single women in Umoja, a number of whom had children, were in fact "outside wives."

At the legal level, the rights and duties of each marriage partner, and even the definitiion of marriage itself, is unclear and often controversial (Kuria, 1984). There are several types of marriage: Christian or church marriage and civil marriage, both of which require monogamy to the extent that a second wife would have to be married under customary law only. In addition there is customary marriage, the details of which vary among the ethnic groups, and Islamic marriage; both of these allow polygyny. The majority of marriages even among the urban middle class are contracted according to custom, even if there is also a church ceremony. The church marriage in effect assures the first wife of at least a superior status in relation to any succeeding co-wives. It is also a sign of economic status. Whatever the marriage type, a wife has very few actual protections against the main risks she faces: lack of support, lack of child support, and physical abuse. A Marriage Reform Bill which would have unified marriage types, made polygyny contingent on the agreement of the first wife, and outlawed the custom of "wife-beating" has twice been voted down in parliament. An Affiliation Bill, which would have enforced child support, was likewise never passed.

In customary marriages, that is, in nearly all marriages, the bridewealth payments continue to define the position of the wife as subordinate. Ninety *percent* of the couples in our sample had paid some bridewealth. Despite the ethnic variations, bridewealth in all cases is conceived as conferring to the husband or his lineage rights to the wife's sexual and domestic labor (uxorial rights) and to her offspring (genetricial rights). It thus conveys rights in women, claims to labor and output, as it were. Bridewealth transactions take place only between those socially defined as men; between the husband and/or his father or other agnates, and the father of the woman or his agnates. Except in the special case of woman-woman marriage in a few groups, women did not and do not own the traditional bridewealth currency, cattle, and so could not receive or give bridewealth. The transition to case payments has made it possible for some women to receive or pay back bridewealth.

Parkin has emphasized important differences between the Luo and other Kenyan ethnic groups in the form and meaning of bridewealth, and hence in the position of women. Among the Luo the payment of valuable bridewealth still secures an unambiguous transfer of all uxorial and genetricial rights in a woman from her natal to her transfer of all uxorial and genetricial rights in a woman from her natal to her husband's descent group. Her incorporation into his descent group is complete, and no distinction is made between uxorial and child payments. Among the Kikuyu, Digo, Giriama and some other groups, on the other hand, the semantic distinction between the two types of payments (which are often made at different times, the child one later), expresses an elment of tentativeness in the marital relationship, allowing it, Parkin argues, to be more easily broken. The status of both wives and children is affected: if the husband or his agnates do not or cannot make the childbirth payments, the children in theory revert to the wife's natal descent group which, traditionally at least, would have been only too glad to receive them.

In case of divorce, the Luo woman returns to her natal group, but has no possibility of taking her children with her. Her relatives repay only part of the bridewealth. Among the Luo the separation and divorce rates are low. The wife's family is often reluctant to support her attempts at independence, since they will have to return part of a large, valuable bridewealth, yet will not receive any of her children. They will have to support a divorcee whose chances of remarrying are slim. Social and economic options are few for the separated or divorced Luo women. Both pressures constrain her toward enduring even quite unhappy marital situations. The potential loss of her children serves as perhaps the most coercive of the various forces of control.

Among the Kikuyu, by contrast, separation and divorce rates seem to be high and on the increase. Bridewealth payments are lower than among the Luo. Since the childbirth payment, the slaughtering of the *ngurario* ram at a ceremonial feast, is so long delayed a part of the total bridewealth transaction, and so small a part of it in terms of value, a separating wife can often take her children with her. Her family is not usually keen on absorbing them nowadays, but she may have the economic wherewithal to set up her own female-headed household. By earning cash in town, she might well be able to pay back the whole of her bridewealth (Parkin, 1980; 1978, pp. 250–282; Nelson, 1978–79). It is clear that there are both socio-economic as well as bridewealth differences which account for the contrast in the positions of Luo and Kikuyu women. Although Kikuyu women's claims on their husband's resources may be less secure because of high separation and divorce rates, on the other hand they live closer to Nairobi and are of the dominant ethnic group. Thus they have more economic options outside marriage than do Luo women.

The persistence of polygyny in some ethnic groups, the persistence of bridewealth generally, the large household size and the prevalence of relatives as guests, together with other factors not discussed, such as the elder generation's continuing control of land inheritance and thus of other rural educational and political advantages, all suggest that the "nuclear family" is at best only weakly emergent among the Kenyan middle class. Yet the forces of change are great. Economically, wealth increasingly depends on salaries, and families are increasingly able to guarantee status and property (e.g. educational advantages) to only a small, rather than a large, family. Strong cultural forces for change operate through the mass media and the education system. The education of women may be particularly important. Educated Kenyan women express a desire for a more emotionally close and egalitarian relationship with their husband, for monogamy, and for husband involvement in childrearing (Whiting, 1977; Parkin, 1978, p. 261; Kayongo-Male and Onyango, 1984, pp. 65–67). To what extent are such aspirations being realized?

Access to Conjugal Resources

The chief economic resource of the urban middle-class family is wage or salary income, initially paid to the individual. This may be supplemented by income from a farm or small business in the rural areas, but the combination of high salaries with significant incomes from rural enterprises is generally available only to the African elite (which is not represented in this sample). The important question becomes, therefore, how are income resources dispersed and controlled within the domestic unit? We examined two aspects of financial relations: (1) the extent to which there was a separated or "autonomous" pattern of income control, as opposed to a "joint" or pooled income pattern; and (2) the extent to which income was actually spent by either the husband or the wife.

The question of "pooling" of resources must be looked at separately for husbands and for wives. Informal conversation suggested, for example, that wives were likely to put a greater proportion of their resources into a household pool than husbands did. One indication of the husband's willingness to pool his resources with his wife is found in the presence or absence of joint bank accounts. Only 22% of the men in the sample reported that they maintained such joint accounts; 29% in Buru Buru 2, and 13% in both Buru Buru 1 and Umoja. Some of these men could also have had individual accounts in addition to the joint accounts. By contrast, in response to the questions as to whether she kept her own earnings separate or put them into a common household fund, while 56% of the women said they kept most of the money separately. 44% of the women said they put most of it into the common fund, this suggests that pooling of resources characterizes about half the *wives* in the sample, separation of resources the other half. Pooling, however, does not necessarily imply equality of control. In response to the question "Who decides what is done with the money you contribute?", 22% of the women reported that they did, 66% said both husband and wife would decide, and 12% said the husband would.

To measure the differing areas of financial responsibility taken by each partner, and to provide another look at the jointness/separateness dimension, the wife's responses to the following question were tabulated: "Between you and your husband, who usually pays for the following items? (Read list.) Would you say it was . . . (read options for answers)." The eight expenditure items were food, children's clothing, wife's clothing, husband's clothing, wages of housegirl, children's school fees, large furniture items, and rent or mortgage (Table 6). Use of the average of these items as an index of financial contribution assumes that these items give a reasonable approximation to the major categories of budgetary outlay in urban African households. The options for answers were five, ranging in an ordinal scale from "husband alone" through "husband and wife equally" to "wife alone." Any movement away from the "husband only" category can be interpreted as an increase in the wife's financial responsibilities and a decline in husband financial dominance. Alternatively, one may collapse the four categories at each of the autonomous ends of the scale, and look simply at the jointness/ separateness dimension.

On the average, husbands tended to be responsible for the larger number of expenditures. Strong wifely financial autonomy was not a very frequent situation, nor was equality in financial contributions.

Patterns of division of responsibility by expenditure area are evident. To a certain extent the very wording of the question forced respondents to make the distinctions between expenditure items and contributors, but the wives and husbands seemed to have no difficulty answering the question or making the distinctions. Wives tended to be responsible for the wages

TABLE 6   DIVISION OF FINANCIAL RESPONSIBILITY   (percentages)

	Food	Child's Clothing	Wife's Clothing	Husband's Clothing
Husband Only	25.7	24.6	27.5	43.3
Husband Mostly	14.9	13.4	15.6	34.9
Husband and Wife Equally	35.1	45.9	28.6	18.8
Wife Mostly	20.1	12.3	20.6	1.9
Wife Only	4.2	3.7	7.7	1.1
	100.0	99.9	100.0	100.0

	Wages-Housegirl	Children's School Fees	Furniture
Husband Only	18.9	42.4	41.7
Husband Mostly	16.0	22.9	26.7
Husband and Wife Equally	23.4	24.7	24.6
Wife Mostly	32.0	6.5	4.2
Wife Only	9.7	3.5	2.8
	100.0	100.0	100.0

	Rent/Mortgage	Average
Husband Only	51.9	34.5
Husband Mostly	28.8	21.7
Husband and Wife Equally	13.9	26.9
Wife Mostly	3.6	12.6
Wife Only	1.8	4.3
	100.0	100.0

of the housegirl, while husbands tended to be responsible for the rent or mortgage, for children's school fees, for large furniture items, for his own clothing, and for a car if there was one. Food, and clothing for children tended to be shared.

To explore the correlates of degree of wifely financial responsibility, a scale called FINRESPW was created from the accumulated, equally weighted

TABLE 7 CORRELATES OF WIFE'S FINANCIAL RESPONSIBILITY
AND DECISION-MAKING POWER

FINRESPW			WIFDEC	
	r			r
EMPLW	.64*		DIFAGE	-.24*
EDUCW	.32*		EDUCW	.22*
DIFAGE	-.28*		EDUCH	.21*
EDUCH	.22*		EMPLW	.18*
HHINC	.22*		HHINC	.16
AGEW	.19*		KIKW	.15
LUOW	-.19*		KIKH	.14
HLAB1	.18*		ONLYW	.13
HSGRL	.16		FINRESPW	.12a
			AGEW	.12a

*significant at the .01 level
All are significant at .05 level unless noted.
a - not significant

scores on budgetary items. It ranged in value from 1 to 5, the higher the score, the greater the degree of wifely financial responsibility. Pearson correlation coefficients between this and a number of other variables are shown in Table 7. FINRESPW was most highly correlated with whether or not the wife was employed. In addition it was weakly but significantly correlated with the wife's education level, the difference in ages of the spouses (negative correlation), the husband's education, household income, the age of the wife, whether or not the wife is Luo (negative correlation), the degree to which the husband does *not* do housework, and the presence of a housegirl. That is, within the population under study, as these factors increase (or decrease), so does the degree of wife's financial responsibility within the household. Simple correlation does not indicate the direction of causation or the interactions among variables. Multiple regression analysis with these variables indicated that wife's employment explained 41% of the total variance in FINRESPW, the other factors each contributing only very small increases in explanatory power. Some of them may be important primarily because they affect employment.

Analysis focusing on the jointness/separateness dimension in this series of questions revealed that jointness, or the "husband and wife equally" response, became more frequent as wife's employment, her education, and household income rose. The responses varied markedly according to estate, the lower-income estate, Umoja, with the highest levels of female unemployment, having the lowest levels of the "joint" response and the highest frequencies of the "husband only" response (Stichter, 1987). This finding is consistent with that of studies in both Lusaka and Lagos, which also found that "pooling" was more common among the middle class, and increased as income and education increased (Munachonga, 1984; Mack, 1978).

But as to the actual extent of income pooling among the middle class, Kenya would seem to offer a contrast to the situation in West Africa. Although comparison is difficult because of sampling differences, Mack (1978, p. 811) found that 3% of the middle class women in her small, non-random Lagos sample pooled resources, but Karanja found in a 1979 random sample of 150 female civil service workers in Lagos that 84–96% said they had full control over their own earnings, and none had joint bank accounts with their husbands (Karanja, 1981, p. 57). Lewis (1977, pp. 180–181) found that only 17% of wives in white-collar occupations in Abidjan pooled incomes. Our figure is 44%. Lewis argues as did Karanja's informants, that the separation of earnings helps the wife maximize her decision-making power within the marriage, although her earnings do not provide an alternative to marriage. By contrast, many Kenyan women, it seems, still find it a struggle to maintain control over their earnings.

Among observers and popular commentators in Kenya, domestic budgeting is acknowledged to be a problem. There is a lack of normative consensus as to how such matters are to be handled. This is not, as some have suggested, because African traditions provide no guides, but rather because African women are demanding more decision-making influence. For example, women protest the widespread assumption that the wife's salary is meant to be spent on the household while the husband's may be spent on entertaining friends or on business enterprises in the rural area (Kayongo-Male and Onyango, 1984, pp. 29–30). They complain, too about the lack of financial trust and openness in marriage, and their lack of control over their own earnings:

> Few African men discuss their earnings with their wives. This means that at times the wife may not even know how much her husband is making. While this is the case, the husband may demand to know or be given her salary. In certain situations the husband buys everything with her money and the wife must be content with his choice. At times part of this money is used to support the husband's family, but the wife cannot use it to support her family. Once the dowry is paid, the wife is not expected to support her family, and whatever she earns in marriage is considered to belong to the husband. This attitude creates problems in the families especially where the wife feels exploited as she has no say about what she earns. (Kayongo-Male and Onyango, 1984, p. 68).

TABLE 8   Spheres Of Family Decision-Making   (percentages)

	Car Purchase	Land-business	Children's Schooling	House-girl	Index (Avg.)
Husband decides	30.0	22.8	13.3	4.2	17.6
Decide together	69.1	74.8	78.8	39.6	65.6
Wife decides	.8	2.4	7.9	56.2	16.8
	99.9	100.0	100.0	100.0	100.0

In general, our findings accord with these observations. Pooling of earnings and other aspects of financial decision-making (see below) remain areas of great domestic conflict. (For evidence of similar conflicts among Zambian middle-class couples, see Schuster, 1979, pp. 118–122; see also Hansen 1984).

Conjugal Decision-Making

Many studies have argued that the greater a woman's financial contribution to a marriage, the greater will be her decision-making power within it. Our findings give only qualified support to this view. A series of questions on decision-making sought to tap this dimension of conjugal relations. The questions were framed in hypothetical terms:

(a) "If you or your husband were to buy a car, how would you decide on that?" (read options)
(b) "If you or your husband were to buy a piece of land or invest in a business, who would usually decide on that?" (read options)
(c) "If you had to decide where to send the children to school, who usually decides on that?" (read options)
(d) "What about if you had to get a new housegirl, who would decide on that?" (read options)

These are typically major decisions for most Kenyan families, but they are of course not exhaustive of all areas of decision-making. To get some approximation to overall family decision-making, the wife's responses to the four questions were either averaged (Table 8) or given equal weight and aggregated to form a scale, WIFDEC. Use of this measure gives greater weight to major as opposed to everyday kinds of family decisions (such as what to have for supper), and to this extent may understate the woman's deciding role.

The WIFDEC scale shown in Table 7 ranged from 1 to 3, the higher the score the greater the wife's decision-making power. A number of factors correlate with an increase in WIFDEC, including the difference in ages of spouses (negative correlation), the wife's education level, the husband's education level, the employment of the wife, household income, whether the wife or husband was Kikuyu, and whether she was the only wife. These correlations are low but significant. Previous analysis indicated that a significant correlation with estate residence also exists. The "husband decides" response in each of the four areas, and averaged over all of them, was more frequent in Umoja than in Buru Buru 1, and in Buru Buru 1 than in Buru Buru 2. His separate role in decision-making, therefore, probably declines as income and wife's employment increase. Though her separate role (the "wife decides" response) conversely increases, this is almost solely because it increases with respect to the hiring of the housegirl; leaving this question aside, it hardly increases at all. Similarly, the "decide together" response increases as one moves from the lower-income to the higher-income estate only if the housegirl question is omitted from the analysis. (Stichter, 1987).

In the sample as a whole (Table 8) there was one significant division of decision-making authority. In hiring the housegirl the wife had undisputed deciding power in the majority of cases. Paying the maid was also usually the province of wives (Table 6). This situation reflects the fact that domestic work is considered to be the woman's responsibility, though she is allowed to hire a substitute. As her income increases, she is likely to do just that. The presence of a housegirl was significantly correlated with the employment of the woman and with household income.

None of the factors examined, including wife's financial contribution to the household (FINRESPW), appear to account in any major way for the variance in wife's decision-making power. Further analysis is needed to explain this outcome. The most likely hypothesis, however, is that it is mainly in the areas of the housegirl, children's schooling, and perhaps other unmeasured areas, that the wife's decision-making increases as her financial contributions do. In the other measured areas, those of car purchase and land or businesses, her decision-making probably does not increase commensurate with the increases in her financial contributions.

It is clear that there is no strong trend evident in this sample of separate or "autonomic" decision-making. The most frequent response was the "decide together" one (Table 8). This contrasts with the Ghanaian civil servant families described by Oppong (1974). Her sample was small but intensively studied. It was, unfortunately, not drawn so as to be representative, but still, out of 61 cases, 19 (33%) could be classified as autonomic, 12 (21%) as husband-dominated, and 27 (46%) as "syncratic" or joint on the decision-making scale (Oppong, 1974, p. 122; 3 cases unclassifiable). This represents a much lower percentage of families with markedly joint patterns of decision-making than was the case in our sample. Oppong also found that as the wife's resources in financial and educational terms rose, there was a trend toward overall jointness. We found this to be true with regard to income pooling, but only weakly so with regard to decision-making. Kenyan middle-class wives appear to be having trouble translating earnings into truly joint or equal decision-making.

Domestic Work

The productive activities of the urban middle-class domestic unit seem small by comparison to those which take place away from the household, in the wage economy. But the familiar picture of the separation of work and family under capitalism, and of the family as a unit solely of consumption, has been overdrawn. Various lines of feminist critique have pointed to the considerable amount of actual production that still takes place within the home. Transforming consumer goods for end-stage consumption, or giving them use-value is, as writers in the domestic labor debate point out, productive work. Maintenance of consumer durables, slowing their rate of deterioration, is work. Consumption itself, as Batya Weinbaum pointed out, can be work: standing in lines to purchase food, arranging transportation, scheduling medical appointments, deciding what to spend when (Weinbaum, 1979). And, most middle class wives are expected to perform what Hanna Papanek has called "status production work": entertaining and in other ways maintaining social ties which enhance the family's claims to status (Papanek, 1979). Finally, the family's central activity of bearing and rearing children must be counted as work.

The actual content of domestic labor, including reproductive labor, is defined by the relationship of the domestic unit to the larger economic context, under capitalism, by its class position. Further, the content is affected by the availability of domestic technology. In the popular magazines it is often asserted that for urban Kenyan families compared to rural ones, the amount of domestic work has greatly decreased, both because of increased commodification (purchasing bottled gas as opposed to gathering firewood) and because of advances in domestic technology (gas stoves). Strictly speaking this comparison cannot be made, because no distinction between domestic and other work can be made for the peasant household. Domestic work is a category defined by the degree of commodification of labor and goods; as capitalism develops domestic work constitutes an ever-declining residual of those tasks and products which have not yet been subsumed under capital. In the rural areas of Kenya a vast amount of agricultural work is still organized through the domestic unit, yet its product may or may not be marketed and it is not accurately called "domestic labor."

That said, it is still interesting to take those delimited tasks which are now considered domestic work in urban conditions and read them back into the rural sector. I would argue that the time middle-class women spend in home maintenance, cooking, shopping and child care has not decreased as much as is commonly thought. In some respects it has actually expanded as a result of the rising living standards. In the area of "housework" or maintenance a great deal of expansion of domestic work has taken place. High cash incomes in the middle class have made possible a materially different life style. High-status possessions such as houses, cars, furniture, appliances, and wardrobes of clothing now have to be first, purchased, and then, maintained—mopped, dusted, washed, ironed, polished, repaired, and protected from theft. All this takes a great deal of time.

Higher incomes have also made possible adoption of new technologies for domestic work which are supposed to save labor. Many of the families in the sample had such amenities as fridges, gas stoves, and running water. However, any time or labor these appliances may save is greatly offset by the rise in the standards of domestic life, by the demands for higher cleanliness levels and greater variety in food and clothing.

It is possible that time spent at food procuring and preparation may have decreased. The urban middle class woman does not spend the long hours hauling water, making charcoal or pounding grain that a rural woman would. But against this has to be weighed shopping time in crowded markets, transport time (traffic is slow in the city, bus service poor to these outlying estates), new varied diets, and the number of dishes and utensils that need washing up after each meal. Food preparation is less physically demanding than in the rural areas, but changing standards do not automatically streamline it.

In the other major area of domestic work, child care, modern values have also increased the amount of time input considered necessary. The middle-class mother often feels she should set aside time specifically for child socialization, particularly to prepare children early for the highly competitive school system. But she has little time for such work; employment cannot be combined with child care as it can in peasant society. And, in a complicated house and neighborhood of strangers, young children can no longer be so safely left in the care of older siblings or nearby kin, even if siblings or kin were available. Part of the solution to these pressures is to have fewer children. Even so, many middle income women express worry about the discrepancy between their ideals of child-rearing and the time they are able to give to it. In particular they worry about the influence of

TABLE 9   DIVISION OF HOUSEHOLD LABOR   (percentages)

	Preparing & Serving Meals	Purchasing Food	Purchasing Large Items
Husband	5.3	34.0	95.5
Wife(s)	89.8	64.1	2.2
Housegirl	3.4	0	0
Children	.4	1.1	1.5
Other	1.1	.8	.7
Total	100.1	100.1	99.1

	Cleaning	Washing	House Repairs	Child Care–Day
Husband	3.9	3.5	60.2	2.7
Wife(s)	77.1	83.4	4.0	40.2
Housegirl	16.3	10.8	0	46.5
Children	1.6	.4	3.3	1.3
Other	1.2	1.9	32.5	9.3
Total	100.1	100.0	100.0	100.0

	Child Care Evening	Purchasing Child's Clothes	Take Children To School	Index
Husband	5.8	75.8	34.7	32.1
Wife(s)	88.4	23.4	13.3	48.6
Housegirl	2.7	0	2.7	8.2
Children	1.8	.8	40.0	5.2
Other	1.3	0	9.3	5.8
Total	100.0	100.0	100.0	99.9

the housegirl on the children, since she is the one who spends the most time with them when they are young.

Who is to do all this domestic work is always subject to a certain amount of negotiation. As Penelope Roberts puts it, there is always a "sexual politics of labor" in the household (Roberts, 1984). Table 9 reports the division of household labor found in our sample. The percentages shown are from the wife's answers to the question "In your family now, how are the household tasks usually done? Who would you say usually does the following . . . (read list of tasks)." The Index is the average percentage for the ten tasks. It thus measures task inputs rather than time inputs, as well as assuming that the ten tasks represent a good approximation to the sum of household work and that each should weigh equally in the summation.

Table 9 shows that despite the help of a housegirl, it is still the wife who performs the most household tasks. The wife is by far the person who most frequently prepares and serves meals, purchases food, does the washing, and takes care of the children in the evenings. The housegirl helps with all of these tasks (as shown by our probe questions, which distinguished between helpers and primary performers), except shopping at which the husband helps. The housegirl's range of tasks, and thus total work contribution, is surprisingly limited. The husband's contribution to household work is by no means negligible according to these measures, since he is active in purchasing children's clothing, purchasing furniture items, making small repairs around the house (although often a repair man or "fundi" is called in), and in some cases usually when he has a car taking children to and from school.

As a measure of total household work contribution, the task index probably understates both the wife's and the housegirl's total labor time and overstates the husband's. The tasks in which the husband is most active are not the daily repetitive chores of meal preparation, washing, cleaning and child care, but rather the less frequently performed tasks of purchasing and making repairs. Considering only the most time-consuming and house-bound tasks, the wife does the bulk of the work.

Munachonga found a similar segregation of household tasks in a 1982 study of 100 couples in Lusaka, Zambia. She points out that purchasing large items and providing for children's educational needs are "new" tasks, and as such are more acceptable to husbands than the traditionally female tasks of cooking and cleaning (Munachonga, 1984).

For correlation purposes a composite scale of husband's work contribution (HLAB1) and wife's work contribution (WLAB1) were created from the wife's answers. (In this area and throughout the study discrepancies were found between the husbands' answers and the wives' answers. These will be the subject of a separate report.) The most important finding was that the husband's work contribution was not highly correlated with any of the major social background variables: age of husband or wife, education of husband or wife, wife's employment, household income, or ethnicity. That is, changes in these variables did not produce changes in the husband's household work contribution. Only one variable did produce significant changes. A positive correlation of $r = .179$ with $p > .01$ was found for residence of the wife. When the wife resides elsewhere, the husband is likely to do more housework.

By contrast, the wife's household work contribution (WLAB1) does decrease as her employment and household income increase. The correlation with EMPLW was $r=.29$, $p>.01$, and with HHINC was $r=.23$, $p>.01$. As income increases, the wife is able to transfer more of the housework to the housegirl.

According to our results, children's work contribution in urban middle class families is not very great (Table 9). Even a specific probe on the questionnaire as to the role of children failed to elicit a greater response here. Male and female children are helpful in taking themselves and other children to school, and female children help with cleaning, child care, repairs and purchases. But they do not usually take the main responsibility of any listed task, except perhaps getting themselves to school. This contrasts greatly with the situation in the contemporary rural areas (Kayongo/Male and Walji, 1978). The low level of child participation is mainly attributable to the competing demands of schooling, but may also be related to the increasing complexity of household tasks.

Other relatives staying in the household are not reported to be very helpful either. Sometimes they might make meals or wash their own clothes, or mind the children. but they do not take primary responsibility for any task, because their stay is thought to be temporary and because they are often occupied with job-hunting. Several wives mentioned that having relatives stay with them increased the household work.

After the wife, the person who most frequently performs household tasks is the housegirl. Her work makes it possible for wives to combine child-rearing and full-time employment. In the sample 79% of those in Buru Buru 2, 78% in Buru Buru 1, and 43% in Umoja had housegirls. Typically, they are poorer uneducated girls from the rural areas, but surprisingly only about a quarter of them were close relatives of their employers. Higher income families, in fact, were less likely to have housegirls who were kin. Housegirls work long hours, almost all day every day, for very little pay. They do basic child care during the day, house cleaning and clothes washing. They help with food preparation and evening child care (Table 9). From the wife's point of view, there are limits to what housegirls can be expected to do. Typically they cannot be given jobs which require handling money (e.g. shopping), going far from the house (e.g. taking children to school), or dealing with modern appliances (making repairs).

The housegirl system poses a number of labor management problems for the wife. The first is finding and retaining a housegirl in a situation of fairly high turnover. This is normally the wife's responsibility, and it may involve a trip to the rural area. If a housegirl leaves, the wife may have to lose a week or so of work finding another one, with the result that her employer will be annoyed and inconvenienced. The second problem is in the quality of work. This is not only a question of theft or carelessness with household valuables; many of the most common concerns revolve around child care. Many women in the sample worried about the quality of child care. For example, a surprising number of housegirls speak a different language from their employer, or a less educated version of the same language. The child learns either the wrong language or poor speaking habits. On the emotional level, the maid may compete with the wife for the children's affections, giving rise to jealousy and anxiety. The nurse may be tempted to give the children forbidden favors in order to protect herself from harmful revelations by the children to the parents (cf. Kayongo-Male and Onyango, 1984, pp. 22–23; Onyango, 1983). In all, there was little indication of a lax attitude by wives toward child-rearing, such as reported by Schuster for Lusaka (Schuster, 1979, pp. 112–113)); rather it seemed that many wives worried about their inability to influence their children or provide for them properly. From the housegirl's point of view, of course, things look different. While many come to town expecting to better themselves and to find husbands, they are often disappointed. After a period of time, most begin to feel exploited by the long hours and low pay and uncomfortable with the subordinate position they have in the household.

Conclusion

Our data, then, do not yield a picture of the independence of middle-income women *from* marriage, or even of great autonomy *within* it. Rather, there is substantial income pooling, but a persistence of separate decision-making areas and a good deal of husband dominance in financial decision-making. Household work remains task-segregated and is still largely the province of wives. Wives rely on housegirls as key helpers. In some ways this pattern of division of household labor is not new. In precolonial Kenyan societies a wife might well have been able to get help from a junior co-wife, a slave girl, or older female children. What has changed is only the mechanism through which some women gain access to other women's labor. Today it is predominantly a wage relationship.

Barbara Lewis has observed that for Ivoirian women of all classes, marriage remains as desirable an asset as it was traditionally (Lewis, 1977). Their control over their own financial resources makes it possible for them to maintain leverage and autonomy within marriage. Audrey Smock, on the other hand, describes Ghanaian middle-class women as having moved from traditional autonomy to modern subordination (Smock, 1977). The traditional spousal decision-making arrangements in Kenya did not accord the wife as great a range of independence as was the case in some West African societies. This consideration might lead us to view contemporary decision making relations in the urban middle-class as progress. The danger in Kenya, however, is that traditional subordination in marriage will persist. The path of change may be different from that described by Smock for Ghana, but the contemporary obstacles are the same.

Notes

1. This research was funded by grants from the Social Science Research Council and the Ford Foundation Program for Women in East and Central Africa. As usual, these organizations bear no responsibility for the interpretations advanced here.

References

Ekejiuba, Felicia. (1984) "Contemporary Households and Major Socioeconomic Transitions in Eastern Nigeria: Towards a Reconceptualization of the Household," paper presented to the Conference on Conceptualizing the Household, Harvard University, Cambridge, MA, November.

Goode, William J. (1964) World Revolution and Family Patterns. New York: The Free Press.

Goody, Jack. (1976) Production and Reproduction: A Comparative Study of the Domestic Domain. Cambridge: Cambridge University Press.

Guyer, Jane. (1981) "Household and Community in African Studies," The African Studies Review 24, 2/3: 87–137.

Guyer, Jane I. and Pauline Peters, eds. (1984) Conceptualizing the Household: Issues of Theory, Method and Application. New York: Social Science Research Council.

Hansen, K. (1984) "Negotiating Sex and Gender in Urban Zambia," Journal of Southern African Studies. 10, 2: 219–238.

Karanja, Wambui Wa (1981) "Women and Work: A Study of Female and Male Attitudes in the Modern Sector of an African Metropolis," pp. 42–66 in H. Ware, ed., Women, Education and Modernization of the Family in West Africa, Department of Demography, Australian National University.

Karanja-Diejomaoh, Wambui (1978) "Disposition of Incomes by Husbands and Wives: An Exploratory Study of Families in Lagos," in C. Oppong, et. al., Marriage, Fertility and Parenthood in West Africa. Department of Demography, Australian National University, Canberra.

Kayongo-Male, Diane and Philista Onyango. (1984) The Sociology of the African Family. London and New York: Longman Group Limited.

Kayongo-Male, Diane and Parveen Walji. (1978) "The Value of Children in Rural Areas: Parents' Perceptions and Actual Labor Contributions of Children in Selected Areas of Kenya." Department of Sociology, University of Nairobi, Seminar Paper No. 27.

Kuria, Gibson Kamau. (1984) "The African or Customary Marriage in Kenya Law Today," paper presented to the Conference on Conceptualizing the Household, Harvard University, Cambridge, Mass.

Lewis, Barbara C. (1977) "Economic Activity and Marriage among Ivoirian Urban Women," pp. 161–191 in A. Schlegel, ed., Sexual Stratification: A Cross-Cultural View. New York: Columbia University Press.

Lloyd, P.C. (1967) Africa in Social Change. London: Penguin.

Mack, Delores E. (1978) "Husbands and Wives in Lagos; The Effects of Socioeconomic Status on the Pattern of Family Living," *Journal of Marriage and the Family* 40, 4 (November): 807–816.

Marks, Shula and Richard Rathbone. (1983) "The History of the Family in Africa: Introduction," *Journal of African History* 24, 2: 145–161.

Medick, Hans and Davin Warren Sabean. (1984) "Interest and emotion in family and kinship studies: a critique of social history and anthropology," pp. 9–27 in H. Medick and D. Sabean, eds., *Interest and Emotion: Essays on the Study of Family and Kinship*. Cambridge: Cambridge University Press.

Munachonga, Monica. (1984) "The Conjugal Power Relationship: An Urban Case Study in Zambia," unpublished paper, Sussex University.

Nelson, Nici. (1978–79) "Female-centred families: changing patterns of marriage and family among buzaa brewers of Mathare Valley," *African Urban Studies* 3 (Winter): 85–103.

Onyango, Philista. (1983) "Working Mother and the Housemaid as a Substitute: Its Complications on the Children," pp. 24–31 in G. Were, ed., *The Underprivileged in Society: Studies on Kenya*. JEARD, Vol. 13.

Oppong, Christine. (1974) *Marriage Among a Matrilineal Elite: A Family Study of Ghanaian Senior Civil Servants*. London: Cambridge University Press.

Papanek, Hanna. (1979) "Family Status Production: The 'Work' and 'Non-Work' of Women," *Signs* 4, 4: 775–781.

Parkin, David. (1978) *The Cultural Definition of Political Response: Lineal Destiny Among the Luo*. New York: Academic Press.

————. (1980) "Kind Bridewealth and Hard Cash: Eventing a Structure," pp. 197–200 in J. Comaroff, ed., *The Meaning of Marriage Payments*. London: Academic Press.

Poster, Mark. (1978) *Critical Theory of the Family*. New York: Seabury Press.

Roberts, Pepe. (1984) "The Sexual Politics of Labor and the Household in Africa," paper presented to the Conference on Conceptualizing the Household, Harvard University, Cambridge, Mass.

Schuster, Ilsa (1979) *New Women of Lusaka*. Palo Alto, Ca.: Mayfield Publishing.

Scott, Joan and Louise Tilly. (1975) "Women's Work and the Family in 19th Century Europe," *Comparative Studies in Society and History*, 17: 36–64.

Sharma, Ursula. (1986) *Women's Work, Class and the Urban Household*. London: Tavistock.

Smock, Audrey. (1977) "Ghana: From Autonomy to Subordination," pp. 173–216 in J. Giele and A. Smock, eds., *Women: Roles and Status in Eight Countries*. New York: John Wiley.

Stichter, Sharon. (1987) "Women and the Family: The Impact of Capitalist Development in Kenya," in M. Schatzberg, ed., *The Political Economy of Kenya*, New York: Praeger.

Tilly, Louise and Joan W. Scott. (1978) *Women, Work, and Family*. New York: Holt, Rinehart and Winston.

Vaughan, Megan. (1983) "Which Family? Problems in the Reconstruction of the History of the Family as an Economic and Cultural Unit," *Journal of African History* 24, 2: 275–283.

Weinbaum, Batya, and Amy Bridges. (1979) "The Other Side of the Paycheck," pp. 190–205 in Z. Eisenstein, ed., *Capitalist Patriarchy and the Case for Socialist Feminism.* New York: Monthly Review Press.

Whitehead, Ann. (1977) "Review Article: J. Goody Production and Reproduction," *Critique of Anthropology* 9/10: 151–159.

―――. (1984) "Beyond the Household? Gender and Kinship-based Resource Allocation in a Ghanaian Domestic Economy," paper presented to the Conference on Conceptualizing the Household, Harvard University, Cambridge, Mass.

Whiting, Beatrice. (1977) "Changing Life Styles in Kenya," *Daedalus,* Spring: 211–225.

10

Trapped Workers: The Case of Domestic Servants in South Africa

Jacklyn Cock

Feminist theory is becoming increasingly sensitive to the complex inter-relations of race, gender and class. The intersection of these three lines of oppression within the experiences of most black women in South Africa raises important questions regarding both the limits and the possibilities of feminist struggle. Feminists in South Africa are forced "to recognize that white women stand in a power relation as oppressors of black women" (Carby 1982, p. 214). This power relation is dramatized in the institution of domestic service, which is the focus of this chapter. However, the chapter is not only concerned to analyze power relations within this institution, but also argues that the domestic labor debate, which has attempted to theorize domestic labor, has been pitched at too high a level of abstraction. The question it poses can only be examined in concrete practices in specific societies. In this sense the institution of domestic service in South Africa yields important theoretical as well as strategic insights.

In South Africa poverty, labor controls and a lack of employment alternatives combine to "trap" about one million black women in domestic service. These women are subject to intense oppression, which is evident in their low wages, long working hours and demeaning treatment by their white female employers.[1] "She does not see me as a woman. She looks down on me." These are the views expressed by many domestic servants, who feel like slaves leading wasted lives which they are powerless to change.

Other Africans also experience their working lives as a form of slavery. This is because Africans in South Africa are one of the most regimented labor forces in the contemporary world. This regimentation is secured through influx control regulations which control the flow of African workers into the so-called white areas which constitute eighty five percent of the total land area in South Africa.[2] No African may remain in a "white" area for more than 72 hours unless he or she complies with certain qualifications. These qualifications are fifteen years continuous residence in the same area, or ten years continuous employment by the same employer, or a one year labor contract. These regulations effectively divide the African population (seventy five percent of the total) into insiders, with rights to seek employment and accommodation in "white" urban areas, and outsiders, who are restricted to the so-called homelands.

Influx control operates very coercively upon African women and binds domestic servants especially tightly. All domestic servants have to be registered. Migrants on one year contracts may have their contracts renewed annually as long as they remain in the same job. But, as with other African migrants, changing jobs requires a new employer to make a special application to a local administration board proving that no local African labor is available. Thus, in effect this legislation imposes an embargo on the entry of unskilled African women into white urban areas, and binds domestic servants to their present employers. Losing employment could well mean forced removal to the teeming rural slums of the homelands.

Unlike other African workers, domestic servants are situated in a legal vacuum within this coercive structure. They are not protected by legislation; no laws stipulate their minimum wages, hours of work or other conditions of service. They lack disability and unemployment insurance, maternity benefits, and paid sick leave, and are vulnerable to instant dismissal by their employers, who often fail to observe the common law provisions. As one domestic observed, "No matter if I work here for one hundred years I can be dismissed for breaking a cup and get nothing. Not even a thank you." Dismissal increasingly means expulsion from town or white owned farms and forcible return to poverty-stricken homelands.

It is in the homelands that the majority (57 percent) of African women are trapped by influx control laws which prohibit their movement to urban white South Africa (Simkins 1983, p. 57). Lack of employment opportunities means some women survive by seasonal work on white owned farms for wages as low as $1.60 a day. For others, such as Rose living in the homeland of Lebowa, land shortage creates a vicious circle of poverty.

> Now we have no ploughing fields. We are dying of hunger. Once the agricultural officers called us together to teach us how to farm, but this never happened again. They told us to buy fertiliser but it cost $14 or more a bag, and us starving people, we have no money.

By comparison, the African women in the urban areas with legal rights to seek employment and obtain accommodation are fortunate. However, even among this group unemployment is high and brings hardships.[3] As one informant commented, "Unemployment brings three difficulties, sickness, starvation and staying without clothes" (Barrett et al. 1985, p. 92). Unemployment is rising sharply in South Africa and many African women in town survive by participating in "informal sector" activities such as hawking, brewing or childminding. In this manner they manage to eke out a precarious existence in a hostile environment.

While waged employment of African women in South Africa has risen in recent years, the expansion of employment opportunities has not eroded women's disadvantages. The increase in employment was especially dramatic between 1973 and 1982 when there was a 51.7 percent increase in the number of black women employed (Favis 1983, p. 5). But these women are mainly located in the service and agricultural sectors in the least skilled, lowest paid and most insecure jobs. The percentage distribution of African women in the economy is as follows: 11 percent of employed African women are in the professional sector (mainly nursing and teaching which accounts for 95 percent of African women employed in this sector); 3 percent of the clerical sector; 5 percent in the sales sector; 17 percent in agricultural production; 13 percent in manufacturing and 50 percent in services (including domestic service) (Favis 1983, p. 7). Many of the African women employed in these sectors feel they have no option but to acquiesce to low wages and appalling working conditions. They are trapped workers, with few alternatives, living out an infinite series of daily frustrations, indignities and denials.

These constraints and deprivations are most obvious among domestic servants, who are coerced into an occupation none of them would choose. The privatized nature of the work, its monotonous and repetitive character, the close control and supervision it often involves, low wages, and the length and irregularity of working hours, were among the reasons cited by informants for domestic service's unpopularity. None of the servants interviewed said they enjoyed their work or derived any sense of fulfilment from it.

You never knock off.
The worst thing about my job is cooking the dog's food and not eating it.
I never sleep at home with my husband and children. Even if I have a half day off I have to come back and sleep here at night.

Many of the servants interviewed were trapped in domestic service by the need to support themselves and their dependent children. Each domestic servant had an average of 5.5 dependents and 58 percent were the sole breadwinners in the sense that no one else in the house was employed in wage labor. Many of these women were single heads of families.

They are often acutely aware of the ironies their work involves. "My madam, she does nothing, but she can live in this nice house and have fat children. My children are hungry" (Barrett et al 1985, p. 34).

This statement points to the most important function of domestic servants in South African society. They provide services which are essential for the reproduction of labor power, both on a daily and a generational basis, for which there are not substitutes provided in a comparably cheap form by either capital or the state. Daily reproduction (the maintenance of the current work force) involves numerous tasks of domestic labor such as cooking meals, washing, mending, cleaning and shopping. Generational reproduction (replacement of the work force) includes child care. Both forms of reproduction, and the role of African domestics in that reproduction, contribute to the prosperity of the South African economy.

However, debate continues over the relation of this labor to capital. Some scholars argue that unpaid domestic labor is productive, or, even if unproductive, produces value because the product of domestic labor is a commodity—labor power. Marx (1976, p. 274) assigns the value of labor power on the basis of the value of the commodities which would be necessary to maintain the health, strength and historically defined standard of living of a worker, i.e. the value of the commodities produced by the worker during necessary labor time. But these commodities are not im-mediately in consumable form when they are purchased with the wage. Additional labor must be performed upon them before they are transformed into regenerated labor power. Consequently some scholars argue that while domestic labor is unproductive, it creates value which is embodied in the commodity labor power. As Wally Seccombe states, when the domestic worker (in this case the housewife)

> acts directly upon wage purchased goods and necessarily alters their form her labor becomes part of the congealed mass of past labour embodied in labour power. The value she creates is realised as one part of the value labour power achieves as a commodity when it is sold. [This] . . . is merely a consistent application of the labor theory of value to the reproduction of labor power itself—namely that all labour produces value when it produces any part of a commodity that achieves equivalence in the market place with other commodities (Seccombe 1974, pp. 8-9).

However, as Paul Smith shows, far from being a mere application of Marx's theory of value, Seccombe's formulation

> represents a serious challenge to it in that it suggests one commodity, labour power, is always sold below its value, since this would be equivalent

to the value of the means of subsistence bought with the wage *plus* the value said to be created by domestic labour (Smith 1978, p. 202).

Smith argues that not all labor produces value; only labor performed within the social relations of commodity production can take the form of socially necessary and abstract labor. Thus Smith insists that domestic labor transfers the value of the means of subsistence to the replenished labor power but does not add to that value.

Nevertheless it is tempting to apply Seccombe's formulation to the second function domestic servants have in South Africa: releasing white women not only for leisure, but also for wage labor. At present 47 percent of the white female population is economically active (Pillay 1985, p. 24). Their wages are considerably higher than those of their domestic servants. Seccombe's analysis might be stretched to argue that the labor of the domestic servant becomes part of the congealed mass of past labor embodied in the labor power of her employer who is also a wage worker. Thus the domestic servant might be said to create value which is realised as one part of the value labor power achieves when sold as a commodity by her employer.

But one of the difficulties with Seccombe's analysis, whether applied to the domestic labor of the housewife or the servant, is that "it conflates the commodity labour power with the person of the worker" (Maconachie 1980, p. 15). Under capitalism workers are not themselves commodities, as is the case under slavery. The product of domestic labor is a living individual who possesses the capacity for labor power, which may or may not be sold as a commodity on the market. The servant (and wife) produce use values which are essential to the reproduction of labor power, but labor power only becomes a commodity by being exchanged on the market. As Smith points out, labor power is produced and reproduced, irrespective of whether or not it is to be exchanged as a commodity on the market. Whether this capacity of women is realized or not depends on the pace of capital accumulation as this affects their role in the industrial reserve army.

Wives form a hidden reservoir of labor power—the employment of domestic servants creates a particular flexibility for capital, which may draw upon this reservoir according to its needs. Servants release their employers for wage labor and the effect is an increase in production under capitalist relations of production and hence an extension of the labor force which produces surplus value. Their role in childcare arguably increases the value of labor power of the employer's children because of the extra training (private schools, universities) that the dual family income allows. Thus domestic servants expand the current work force, both directly and indirectly, and increase the value of labor power of some members of the future work force. The first function depends entirely on the pace of capital accumulation.

Domestic service performs other functions as well. It absorbs large numbers of mainly unskilled black workers who would otherwise lack jobs in the present situation of growing structural unemployment. It thus takes up some of the surplus labor power of those who cannot gain wage employment in the dominant levels of the economy. It could also be viewed as a category of disguised underemployment in that earnings are abnormally low, and of visible underemployment in the case of domestic servants who are involuntarily restricted to part-time jobs.

The institution of domestic service serves an ideological function which operates in two opposing directions. On the one hand it socializes whites into the dominant ideological order. Often it is the most significant interracial contact whites encounter, and they experience this relationship in extremely asymmetrical terms. Many white South African children learn the attitudes and styles of racial domination from domestic relationships with servants, particularly nannies. It might be thought that servants are similarly socialized into subordination, and in this sense domestic service would reinforce existing class relations. Certainly domestics are subject to numerous practices and rituals of inferiority. However the servants' response to these practices is the precise opposite of what might be expected. Servants, as well as their employers, are both in an extremely dependent position; this dependence is secured through the state and through an ideology of subordination to which wives and servants respond very differently.

The fulcrum of the relationship between employer and servant is their mutual dependence. Employers depend upon their servants' labor, but with unemployment rising most sharply among black women, an individual servant is easily replaced. Servants depend upon their employers for most of the necessities of life to support themselves and their children. As we have seen, this dependence is secured through the state operated system of influx control.

At the same time, white women, who are the employers of black domestics, are also domestic workers. Even where servants are employed to perform the majority of manual tasks, the sexual division of labor within the home lays most administrative responsibility for household consumption and organization on the wife. In the Eastern Cape study the majority of married respondents said their husbands never helped them with domestic work. Some resented this. "He should occasionally cut himself a slice of bread, or pull up a chair." But the employment of servants reinforces the exclusion of men from domestic labor. "Men should help if there are no servants. On farms here we usually have lots of servants, so it's not necessary." In addition to their responsibility for domestic labor, both wives and servants are isolated in the privatized sphere of the home and have a subordinate status within it. Both are subject to extensive control which involves a submission to personal authority. As servants are bound to their employers through labor contracts and influx control, wives are bound to their husbands through marriage contracts.

The marriage bond subordinates women, securing their economic, legal and sexual dependence. Married women are not treated as autonomous individuals by South African law and social policy. Instead they are defined in terms of marriage, a unit headed by the husband with the wife as dependent. Furthermore in South African law a husband cannot be convicted of raping his wife. By implication, she is a form of sexual property to which he has the right of access.

Many working class housewives are separated not only from the means of production, but also from the means of exchange. They therefore depend upon the redistribution of their husband's wage which is conducted in private between them, and over which they may have little control. Their position thus involves an economic dependence. Over a third of the married women interviewed in the Eastern Cape study did not know their husband's incomes. When I asked how family finances were arranged, I was told:

> I've never been able to find out his income. I have to ask him for money.
> He gives me pocket money when he feels like it.

The fact that the cash income of both wives and servants is often given as "pocket money" underlines the economic dependence of both types of domestic workers.

Such economic dependence promotes deference from wives. The relationship between husband and wife is deferential in that it is hierarchic, traditionally-legitimated and embedded in a system of power (Bell & Newby 1976, p. 164). Most of the white middle class wives I interviewed accepted the relations of male domination and female dependence as natural and inevitable. For instance,

> A woman is a womb—her primary function is to be a good mother.
> Our submission to men is God's law.
> The men should be the head of the house.

Clearly, the dependence and control of wives is not only mediated through the state, but also through an internalized ideology of subordination.

These deferential wives also accept the relations of racial domination as natural and inevitable. Sixty eight percent of those interviewed regarded blacks as indubitably inferior.

> They're stupid and irresponsible . . . in short, very raw.
> They've just come out of the trees.
> Putting them into European clothes doesn't make them civilized.

The common equation of blacks with children provides an ideological space for turning middle aged domestic servants into family dependents. Thirty percent of the employers interviewed described their servants as "one of the family." They are widely viewed as "loyal, obedient and deferential" workers who accept the legitimacy of their own subordination in the social order, and defer to their "natural" superiors.

Yet, research in the Eastern Cape suggests that the deference attributed to domestic servants is more apparent than real. A deferential mask is deliberately cultivated to hide the worker's real feelings. It is a protective device generated by powerlessness and the inability to express overt dissatisfaction.

In the work place the disparity in income and lifestyle between worker and employer is highly visible. The work situation acts as a model of the wider society in the minds of many workers; the inequalities of power and wealth experienced at work reflect more general inequalities in society. Most domestic servants interviewed reject the legitimacy of these differences. The great majority resented the gap between their living standards and those of their employers.

> It makes me angry to look at their gardens and the food they buy for their dogs. It is better than what they buy for us. And the dogs eat off their dishes but we don't.

All showed a sense of relative deprivation and thought they should be paid at least twice their present wage.

> Because I work hard . . . I look after the house and even the dogs, cats and chickens. I have to sort the eggs very carefully and check if they are first grade. (This woman was earning $15 a month in 1979).

All thought that blacks and women were unfairly treated in South Africa.

> Our men treat us badly. Our marriages end like paper fires.
> The whites are standing on our necks with their boots.

However, some seemed to have a sense of personal superiority to whites:

> We are more capable than whites. That is why they try by all means to keep us under their feet.
> You can put a black person in the forest and just leave water with him or her. We can manage because there is a lot we can do.

Dependence on white employers was not transformed into a sense of collective weakness. Perhaps the employers' dependence on their domestics' labor reinforced a sense of domestic workers' own capabilities. Several comments emphasized the helplessness and weaknesses of employers:

> She is lazy. She sits a lot on the *stoep* outside while I have to rush around. She couldn't manage without a slave like me.

Perhaps this is a device for maintaining a sense of dignity in a demeaning role—a subtle inversion of the asymmetrical nature of the relationship. All agreed that domestic servants as a group are badly treated, and some consciousness of a community of interest emerged here. "We are all singing one song. We need the same help with low wages and bad treatment." Fundamental change was seen as inevitable, but, as one woman expressed it, "it will take time. It is not easy to take a piece of meat out of your mouth and share it."

The question must then be posed: Does the institution of domestic service contain tensions in South Africa by promoting deference relationships which exact feelings of loyalty and gratitude? If this were the case, domestic service could be said to undermine solidarity among the oppressed by linking them as individuals to their oppressors. It might then afford a fragile bridge across the contradictions of a society based on racism and exploitation.

At the same time the fact that domestic servants are atomized workers in scattered workplaces means that they are largely excluded from the strikes and stayaways that have been dramatic expressions of mass action by black workers in recent years. The national organiser of the South African Domestic Workers' Association (SADWA) believes that "strikes of domestic workers are two to three years away."[4] She reported that many of SADWA's 20,000 members are "fearful of politics." As a result their participation in the 1985 consumer boycott of white owned shops came into question.[5] Some employers shopped for their domestic servants who were afraid to be apprehended breaking the boycott, or to avoid the high prices charged by black shopkeepers taking advantage of the boycott (*The Star* 11 August 1985, 19 August 1985).

Clearly the relation between wives and servants, or unwaged and waged domestic workers in South Africa encapsulates a mass of contradictions. Further tension is engendered by the fact that many black women fuse these roles, and are engaged in domestic labor in two households—their own and their employers. Available evidence indicates that black men do very little domestic labor in the urban household (Cock et al. 1983, p. 41). The present sexual division of labor within the household seems to be widely viewed by black men as natural and immutable.

I married my wife so that she can give birth to my children, look after
the house, wash my clothes, prepare my meals, and that is all—it ends there
(Van der Vliet 1982, p. 184) (p. 540f CC report long done).

Those servants who are also wives are in an extremely dependent and
vulnerable position. Black women married by a marriage officer without an
antenuptial contract are automatically married out of community of property,
but the husband retains the marital power. The husband acquires guardianship
of the wife, who is a legal minor unable to enter into any binding contract,
even a hire purchase agreement, or credit account, without the prior
permission of her husband.

So far it has been suggested that the institution of domestic service is
of considerable significance in the political economy of South Africa, that
it not only reflects inequalities, but reinforces them in a contradictory way.
However the large-scale employment of domestic servants within white
households is anomalous in two senses.

First, domestic servants are anomalous in an industrial society. Katzman,
in his analysis of domestic service in the USA between 1870 and 1920
characterizes it as a non-industrial rather than a preindustrial occupation.
The occupation has a number of characteristics which differentiate domestic
servants from other wage workers. Other wage workers sell their labor
power as a commodity for a definite period of time in exchange for a money
wage. Work relationships are impersonal and involve a clear separation
between workplace and home, both in temporal and spatial terms. The
domestic servant by contrast frequently works irregular hours, she receives
part of her payment in kind and the live-in domestic servant lives at the
workplace. Employer control often extends into the servant's private life—
for example the regulation of visitors and the inspection of servants' rooms
and goods. The highly personalized nature of the servant's relationship
with her employer and the low level of specialization in domestic roles are
both anomalous in a modern industrial society moving towards specialized
and impersonal work relationships.

Second, the large-scale employment of domestic servants is anomalous
in capitalist society. While the work they do in the maintenance and
reproduction of labor is essential to capital, their employment as wage
workers is not. Braverman writes,

. . . the multitude of personal servants was, in the early period of capitalism,
both a heritage of feudal and semi-feudal relations in the form of a vast
employment furnished by the landowning aristocracy, and a reflection of the
riches created by the Industrial Revolution in the form of similar employment
furnished by capitalists and the upper middle class (Braverman 1974, p. 363).

The number of domestic servants in South African is similarly a heritage of feudal relations and is a reflection of the high standard of living enjoyed by most whites. But in South Africa even white working class households can often afford to employ black women as domestic servants.

There are two ways of looking at this. One can argue that cheap, black domestic labor subsidizes the white working class in South Africa, enabling their necessary means of subsistence to be cheaper than it would be if creches and day nurseries were provided by the state, or if commodities had to be purchased in an immediately consumable form within the capitalist sector. Thus the domestic servant's labor cheapens the cost, for capital, of maintaining and reproducing white labor power. Domestic servants would thus perform the same function as the informal sector arguably has for the black working class. On the other hand, one could argue that the widespread employment of domestic servants by the white working class made the necessary means of subsistence of the white working class more expensive than it would be if the housewife was solely responsible for domestic work. Thus the price of white working class labor power is increased and capital's profits are correspondingly lowered.

The question at issue is how the white workers' necessary means of subsistence comes to be defined. Since the earliest colonial penetration this has included the employment of cheap black labor. Van Onselen argues that because of the particular form colonial domination has taken in South Africa, "The white proletariat built the price of a black servant into the cost of reproducing itself" (Van Onselen 1978, p. 21). This economic concession to white workers has been a relatively small burden to capital, both because the wages of servants are so low, and because capital is largely dependent on the exploitation of black workers.

This illustrates the point that there is no invariant relation between domestic labor and the value of labor power. Molyneux (1979) has pointed out that it is only possible for women to remain in the home as housewives where the value of labor power of the male worker is high enough to cover the cost of maintaining the entire family. By the same terms it is only possible for the wife to employ a substitute—in the form of a domestic servant—where the value of white working class labor power is high enough to maintain the enlarged household. In the same way that the ability of a section of the working class to maintain a wife at home has come to represent a particular index of working class power, the ability of the white working class to maintain a servant is an index of its privileged position in South Africa.

Thus domestic service does not only persist for economic reasons. Research done in the Eastern Cape found that many white households had both servants and labor saving devices. Several employers were reluctant to allow their servants to use these—a pattern that was also reported by Shindler

(1980) in her Johannesburg study. This research underlines the non-economic reasons for employing servants. Servants are an important component of the social display of dominant class lifestyles.

Several employers in the Eastern Cape cited other reasons for employing servants as well. For example, "security"; "I don't feel safe on my own during the day"; or to overcome a sense of social isolation, "I feel very alone in the world when the servants go off in the evening". This illustrates an important aspect of the domestic servant's situation. Carby has made the cogent point that "in concentrating solely upon the isolated position of white women in the western nuclear family structure, feminist theory has necessarily neglected the very strong female support networks that exist in many black sex/gender systems" (Carby 1982, p. 230). In the South African context these female support networks function as strategies of survival. They are of inestimable value to the working class African women coping with the strain of their dual roles as mothers and as workers. A peculiarly South African twist comes from the fact that it is often, as in this case, the black woman employed as domestic servant who dilutes the isolation of the white housewife. Thus domestic servants are squeezed between two households, their own and their employers. Their subordinate status as servants and the long working hours exacted by their employers means that they are full members of neither.

In addition to lessening their employer's sense of social isolation at the cost of exacerbating their own, many domestic servants take considerable responsibility for the care of their employer's children.

> She gets the children up in the morning, gives them their breakfast, walks the youngest to nursery school, has our lunch ready for us when we return.

This responsibility for child care involves one of the central contradictions in the institution of domestic servants. Several servants interviewed stressed that they had to look after two families and neglect their own in the process.

> We leave our children early in the morning to look after other women's families and still they don't appreciate us.
> We have to leave our children and look after our madam's children. We have not time to look after them even when they are sick.

It is black women who suffer most from the neglect of creches by the state. Furthermore, black women generally and domestic servants specifically are most vulnerable to dismissal on the grounds of pregnancy (Cock et al. 1983). One respondent said that the employment of domestic servants explained "why white people's children don't grow up criminals. It is not from having everything they need, but having nannies who watch them every minute of the day and instill discipline." Often the person looking after the servant's children is a daughter who is kept out of school to do so. This perpetuates a vicous circle of poverty, inadequate child care and interrupted education among blacks' children (especially females) while white children benefit from the attention of two mothers.

Molyneux has emphasized that it is the work of child care which "is of the most benefit to the capitalist state" (Molyneux 1979, p. 25). Child care is expensive if it emphasizes child development rather than custodial care. Therefore in advanced capitalist societies "the only large scale possibility that could bring about the socialisation of child care would be for the state to expand its provision" (CSE 1975, p. 14). But state organised institutions for the reproduction of labor power are financed by state expropriation of surplus value. Thus since the state provision of childcare centers, kindergartens and creches would add to capital's costs for reproducing the labor force, this would only be likely to occur in a period of rapid capital accumulation and consequent increased productivity. In such a situation capital would gain from releasing women for wage labor, because that would expand the labor force producing surplus value. But in South Africa the availability of cheap, black domestic labor creates this flexibility, and women can easily be incorporated or expelled from the labor force according to the pace of capital accumulation. Hence this is not a demand likely to be made on the state by the white working class.

The availability of cheap black domestic labor also goes some way towards explaining the absence of strong middle class feminism in South Africa. The employment of domestic servants has softened the tension experienced by many women in advanced capitalist societies between their roles in social production as wage workers and their roles in reproduction as domestic workers. That tension has certainly contributed to feminist demands in most advanced capitalist societies, and its absence in South Africa has no doubt inhibited middle class feminists.

The increasing militarization of South African society could change this. In an escalation of conflict, increasing racial distrust may make white women reluctant to employ black servants, and eager to demand that state provide creches, day care centres and nursery schools. This would both release them for wage labor and extend the state's control over children's early socialization.

Increasing militarization indicates the present crisis of social relations in South Africa. In this crisis black women have to shoulder most of the burden of the reproduction of labor power—in the homelands, in their own homes and as domestic servants in the homes of the dominant classes. In this last role black women may best be described as "trapped workers"; workers with few alternatives, living out a daily round of frustrations, indignities and denials. This chapter documents these denials—the denial of family and social life, of reasonable wages and working conditions, of job satisfaction and security. It also points to the benefits reaped by employers of cheap, black domestic labor, and the contribution of the institution of domestic service towards maintaining relations of inequality in South Africa at present. Domestic servants provide services which are essential for the reproduction of labor power, both on a daily and, (most importantly), on a generational basis. No substitutes are provided in a comparably cheap form by either capital or the state. The role of black domestic servants in this reproduction provides the key to understanding the persistence of the institution of domestic service in South Africa. The deprivation imposed on these black women and their children must be rated one of the more onerous hidden costs of the apartheid system.

Notes

1. Their wages and working conditions are described in Cock (1980). The main data source for this study was a random sample survey that involved 225 interviews with domestic servants and their employers.

2. At the time of writing (March 1985) there is government talk of abolishing influx control. In 1985 the President's Council suggested replacing influx control by a policy of "orderly urbanization," involving the use of the Squatting Act, the Slums Act and zoning and health regulations. However, there is no talk of repealing the Group Areas Act which prevents African people from living outside their prescribed areas. To many observers therefore "orderly urbanization" is merely a label for a modernized form of influx control.

3. The percentage of South African unemployed in 1983 in the white areas was males 4. 1; females 12. 3. In the homelands it was 11. 9 among males and 19. 4 among women (Pillay 1985).

4. Interview by author March, 1985.

5. The consumer boycott of white owned shops which was launched in 1985 is one of the most important strategies for change to emerge within South Africa in recent years. It was a national consumer boycott in protest against the declaration of a state of emergency and had four demands: the lifting of the state of emergency, the removal of the police and the army from the townships, the release of all detainees (over 10,000 people were detained during the year) and political rights for all South Africa's people. At the time of writing none of these demands had been met, but the boycott mobilized African women around the country.

References

Barrett, J., A. Dawber, B. Klugman, I. Obery, J. Schindler and J. Yawitch, (1985) *Vukani Makhosikazi. South African Women Speak.* London: Catholic Institute for International Relations.

Braverman, H. (1974) *Labour and Monopoly Capital. The Degradation of Work in the Twentieth Century.* New York: Monthly Review Press.

Bell, C. and H. Newby (1976) "Husbands and wives: the dynamics of the deferential dialectic," in D.L. Barker and S. Allen, eds., *Dependence and Exploitation in Work and Marriage.* London: Longman.

Carby, H. (1982) "White Women Listen! Black Feminism and the Boundaries of Sisterhood," pp. 212–235 in Centre of Contemporary Studies, *The Empire Strikes Back: Race and Racism in Britain.* London: Hutchinson.

Cock, J. (1980) *Maids and Madams: A study in the politics of exploitation.* Johannesburg: Ravan Press.

Cock, J., E. Emdon and B. Klugman (1983) *Child Care and the Working Mother: A Sociological investigation of a sample of urban African women in South Africa.* Cape Town: Southern African Labour and Development Research Unit.

Favis, M. (1983) *Black women in the South African economy.* Durban: unpublished paper.

Katzman, D. (1978) *Seven Days a Week. Women and Domestic Service in Industrializing America.* New York: Oxford University Press.

Lawton, L. (1985) *Working Women.* Johannesburg: Ravan Press.

Marx, K. (1976) *Capital.* Volume 1. Harmondsworth: Pelican.

Pillay, P. (1985) "Women in Employment: some important trends and issues," *Social Dynamics* 11, 2 (December): 31–38.

Seccombe, W. (1974) "The Housewife and her Labour under Capitalism." *Socialist Woman Special.* London; I.M.G. Publications.

Shindler, J. (1980) "Labour Saving Appliances and Domestic Service," *Southern African Labour Bulletin* 6, 1.

Simkins, C. (1983) "The distribution of the African population of South Africa by Age, Sex and Region-Type 1950–1980," in C. Simkins, ed., *Four Essays on the Past, Present and Possible Future of the Distribution of the Black Population of South Africa.* Cape Town: Southern African Labour and Development Research Unit.

Smith, P. (1978) "Domestic Labour and Marx's Theory of Value," pp. 198–220 in A. Kuhn and A. Wolpe, *Feminism and Materialism.* London: Routledge and Kegan Paul.

Van Der Vliet, V. (1982) "Black Marriage: expectations and aspirations in an urban environment." M.A. thesis. Johannesburg: University of the Witwatersrand.

Van Onselen, C. (1982) *Studies in the Social and Economic History of the Witwatersrand 1886–1914.* Vol. 2. London: Longman.

Notes on Contributors

Jacklyn Cock is a Senior Lecturer at the University of Witwatersrand, Johannesburg where she teaches courses in the Sociology of Militarization, Medical Sociology, Women's Studies and Social Theory. She is presently involved in research on the militarization of South African society. Her best known publication is the book *Maids and Madams, a study in the politics of exploitation* (Johannesburg: Ravan Press, 1980). This generated a great deal of controversy culminating in a dynamite attack on her home from which she was fortunate to escape alive.

Nancy Folbre teaches economics at the University of Massachusetts. She has pursued research on the economics of household production and childrearing through fieldwork in Mexico, Kenya, and Zimbabwe, as well as through historical studies of early New England. Her theoretical interests include the interface between contemporary Marxist and feminist theory, the political economy of childbearing, and public policies toward mothers. She is currently at work on a book entitled *Patriarchal Capitalism in the U.S.*

Jeanne Koopman Henn teaches economic development at Northeastern University in Boston. She pursued research on rural labor and incomes in Cameroon and Tanzania from 1974 to 1981, taught at the University of Dar es Salaam, and recently completed four months of village research with farmers in Bilik Bindik and Mgbaba II in southern Cameroon.

Janet MacGaffey is an anthropologist. She has taught at Haverford, Bryn Mawr and other colleges in the Philadelphia area, and contributed articles on Zaire to scholarly publications. She has just completed a book: *Entrepreneurs and Parasites: the Struggle for Indigenous Capitalism in Zaire*, to be published by Cambridge University Press. She is currently engaged in research on the underground economy in Philadelphia with the Institute for the Study of Human Issues, Philadelphia, and has served as a consultant for the World Bank.

Jane L. Parpart is Associate Professor of History at Dalhousie University in Halifax, Canada. She has written extensively on labor and women in Africa and is the author of *Labor and Capital on the African Copperbelt* (Philadelphia: Temple University Press, 1983). She is currently working on a study of women in the colonial Copperbelt towns of Zambia.

Penelope A. Roberts teaches in the sociology department at the University of Liverpool. She has published widely on Africa. Her article "Feminism in Africa, Feminism and Africa" (*Review of African Political Economy*, 1984) is a landmark in the study of feminism in Africa.

Sharon B. Stichter is Professor of Sociology at the University of Massachusets at Boston. Her publications include *Migrant Labor in Kenya: Capitalism and African Response, 1895–1975* (Longman, 1982), *Migrant Laborers* (Cambridge University Press, 1985), and *African Women South of the Sahara* (Longman, 1984) coedited with J. Hay. She is currently working on a study of women, employment and the family in Nairobi.

Jane Vock is currently a doctoral candidate at McMaster University, Hamilton Ontario, Canada. She recently completed her field research in Zambia for a thesis on the political economy of fertility decision.

Luise White teaches African history at the University of Minnesota. Her book on the history of prostitution in Nairobi is published by Cambridge University Press. She is the author of "Prostitutes, Reformers, and Historians" (*Criminal Justice History*, 1985) and "Prostitution, Identity, and Class Consciousness in Nairobi During World War II" (*Signs*, 1986).

Index

and World War II, 150-152, 155
  See also Domestic service; Labor

Race relations, 211, 212
Radical feminism, 2, 6. *See also*
  Feminism
Repatriation, of town women, 121, 123.
  See also Copperbelt; Law
Reproduction, 2-6, 13, 23
  biological, 4-6, 8-11, 82-84, 87-90,
    178, 195
  control over, 5-6, 10-12, 38
  definition, 7-8, 82
  and domestic service, 21-22
  generational, 8
  human, 9, 41
  of labor forces, 6, 8-10, 18, 21, 208-
    209, 217
  mode of human reproduction, 12-13,
    27
  and prostitution, 18
  relation to capital, 208
  relation to production, 12-15, 23
  social relations of, 6, 11, 13, 82-83,
    86-88
  technology, 11, 13
  See also Childbearing; Child rearing;
    Children; Contraceptives; Family;
    Labor
Reserve army of labor, 8, 20. *See also*
  Labor; Wife
Resistance (of women)
  to patriarchy, 46, 48, 49, 66, 91, 116,
    123, 133
  through second economy, 169

Sacks, Karen, 3
Seccombe, Wally, 6, 7, 12, 53, 93, 208,
  209
Seclusion (*purdah*), 4
Second economy, 19, 162, 168-172
  and state, 168-169
  and women, 162, 164, 166
  See also Informal sector; *Magendo*
Semi-proletarianization, 8
Separate purse, 102, 166

joint purse, 103
  See also Income pooling
Servants, 21-22, 185, 198-201, 205-218
  child care, 200
  creation of labor power, 21-22
  in industrial society, 214
  neglect own children, 216-217
  and prostitution, 147
  release wife's labor power, 209
  security, 216
  as social display, 216
  subsidize white workers, 215
  See also Domestic labor; Domestic
    service
Settlers (Rhodesian), 67, 71-73, 77
  and African chiefs, 73
  See also Kenya; Zimbabwe
Sex/gender system, 2, 11
Sexual intercourse, 10, 18
Sexuality, in bargaining with men, 171-
  172
Sexual politics, 20
  of labor, 16, 197-198
Shona, 15, 63-76. *See also* Zimbabwe
Sierra Leone, 110
Slaves, 16
  domestic, 103
  labor of, 109-112, 209
  See also Labor; Pawns
Social formation, patriarchal tributary,
  67
Socialist feminism, 4, 23. *See also*
  Marxist-feminism
Societies
  matrilineal, 108, 110, 115-118
  patrilineal, 5, 47, 115-118
  post-colonial, 162. *See also* State
  precapitalist, 29
  precolonial, 28, 42, 47, 62, 89
  stateless, 45-46
South Africa, 20-22, 74, 205-218
State
  and child care, 217
  colonial, 48, 49, 162, 168
  and informal economy, 174
  post-colonial, 19, 162, 168
  women's avoidance of, 169

urban centers, 19
use of earnings, 167
as tribute, 66
See also Family; Wife
Workers
and servants, 215
waged, 14, 40–42
waged women, 50
See also Labor; Wages

Working class, 22. *See also* Class;
  Middle class
World systems, 14

Yoruba, 49, 100, 106

Zaire, 19, 161–175
Zambia, 17, 115–138, 198
Zimbabwe, 15, 62–80